NEUROSCIENCE RESEARCH PROGRESS

SYNAPTIC PLASTICITY

ROLES, RESEARCH AND INSIGHTS

NEUROSCIENCE RESEARCH PROGRESS

Additional books in this series can be found on Nova's website under the Series tab.

Additional e-books in this series can be found on Nova's website under the e-Books tab.

NEUROSCIENCE RESEARCH PROGRESS

SYNAPTIC PLASTICITY

ROLES, RESEARCH AND INSIGHTS

SRABONI CHAUDHURY
EDITOR

nova
Medical & Health

Library of Congress Cataloging-in-Publication Data

ISBN: 978-1-53613-606-7

Published by Nova Science Publishers, Inc. † New York

CONTENTS

PREFACE

Connectivity of neurons is genetically hardwired in the different regions of brain and it involves regular reorganisation in response to stimuli. This process of refining of neuronal connectivity to adapt is defined as synaptic plasticity. Synaptic plasticity regulates how neurons interact with each other to execute a physiological response and determine brain ability to change in response to external stimuli. Over the years, studies on the pre- and post-synaptic mechanisms of synaptic plasticity have identified the role of synaptic release and recycling, activity-dependent neuronal activation, stimulus-induced gene expression and neurotransmitter receptor trafficking in neuron rewiring or synaptic plasticity. Lot of efforts has been made to understand the synaptic plasticity in several brain functions. The role of synaptic plasticity has been studied in relation to environmental stimuli as well as during various disease condition. The neurodevelopmental disorders including autism, fragile-x syndrome has been shown to be regulated by synaptic plasticity. The manifestation of various psychiatric illness effects the neural plasticity in several region of brain. To refocus on the idea of understanding the role of synaptic plasticity with cutting edge technology and further reinforcing our understanding of synaptic plasticity in many brain functions, this book will give you an insight in activity-dependent synaptic plasticity as well the role plasticity plays in various diseased condition in light of new research and advanced technology.

Chapter 1 will explain the reader the basis of synaptic plasticity and mechanisms of the synaptic plasticity in interneurons. Chapter 2 gives an overview through the interactions of AMPA and NMDA receptors with integrated mathematical model of synaptic plasticity induction by postsynaptic Ca2+ influx. Chapter 3 provides an insight in neuronal plasticity and experience-dependent behaviors in fruit fly. Chapter 4 explores the synaptic plasticity and electrical adaptation in neurons in relation to psychiatric disorder. Chapter 5 demonstrates the role of synaptic plasticity in autism spectrum disorder. Chapter 6 reveals how hippocampal synaptic plasticity is regulated in fragile-x syndrome. Chapter 7 exhibits the role of neurotrophic factors in regulating synaptic plasticity in psychiatric disorder. Chapter 8 displays the hippocampal synaptic plasticity in context of epilepsy. Chapter 9 unveils the neuronal activity and synaptic plasticity underlying hippocampal memory engram. Chapter 10 unfolds the tools and techniques important to analyze synaptic plasticity.

In: Synaptic Plasticity
Editor: Sraboni Chaudhury
ISBN: 978-1-53613-606-7

Chapter 1

EXPERIENCE-DEPENDENT SYNAPTIC PLASTICITY OF INHIBITORY INTERNEURONS

Vikram Sharma
Molecular and Behavioral Neuroscience Institute,
University of Michigan, Ann Arbor, MI, US

ABSTRACT

Inhibitory circuits are diverse while inhibitory neurotransmission is dynamic in nature and facilitates a wide range of functions in the brain which includes dendritic integration, control of neural excitability, circuit reorganization, and fine-scale refinement of network activity. Synaptic plasticity has a critical role to play in modulating inhibitory circuits. Different brain regions, such as the somatosensory cortex and the auditory cortex, have also been used to study different forms of experience-dependent plasticity. During development and early post- natal life there are critical periods of plasticity during which the neuronal circuitry is most receptive to environmental stimuli. This early experience-dependent activity translates to a more permanent and sophisticated neuronal connection in the adult brain systems. Multiple studies have pointed to the development of inhibitory circuitry as one of the key factors for the onset of critical periods which regulate the synaptic plasticity. In this chapter, we will discuss the role of activity in regulating inhibitory circuits, thereby modifying synaptic plasticity which is required for various brain activity to be performed in a coordinated fashion.

Keywords: GABA, synaptic plasticity, glutamate, synapse

1. INTRODUCTION

The brain shows an enormous ability to adapt to the altered environment. Brain circuits are very receptive to stimuli and is important to coordinate various stimuli to influence neuronal connectivity especially with the inhibitory circuits. The way experience influences the structure and function of neuronal connections, referred to as experience-dependent plasticity which alters the brain wiring. The root of this adaptation is the formation and refinement of neural circuits, allowing our brains to develop, acquire knowledge, learn new skills, and recover from injuries. It is proposed that modification in synaptic plasticity helps the neural circuits to learn and perform complex dynamic tasks. Synaptic plasticity is considered to the basis for many brain functions including pain, learning, memory, emotion, cognition, and so on (Horlemann and Zieglgänsberger, 2009). Synaptic plasticity could be controlled by several signaling cascades which regulates or co-regulates, the efficacy of neurotransmitter (GABAergic and glutamatergic) release. Alteration in the signaling pathway via neurotransmission lead to structural changes in synapse. Input and output differences at the synapse is govern by specific subcellular targeting (e.g., dendrites, soma, axon-initial segment) of postsynaptic cells by different GABAergic neurons. Activity-depend interactions between GABAergic and glutamatergic synapses is thus an important phenomenon for many functional processes involving circuit maturation and fine-scale refinement of network. Hence, several neurodevelopmental and neuropsychiatric disorders implicate both excitatory and inhibitory neurotransmission systems for understanding the disease (Rubenstein and Merzenich, 2003; Nelson and Valakh, 2015; Mullins et al., 2016). In this review we will mainly focus on inhibitory circuits playing role under the influence of activity-dependent synaptic plasticity.

2. Basis of Synaptic Plasticity and Role in Interneurons

The flow of information in a neural network in the brain involves effective communication between the pre- and post-synaptic neurons using a variety of neurotransmitter and neurotransmitter receptors generating electrical or chemical responses (Bear et al., 2007). Strength or the efficiency of synaptic transmission is highly regulated on a timescale of few seconds, hours or even longer leading to an overall change in the neuron network activity and subsequent behavioral modification. In this context, mode of communication between the neurons can be either driven by electrical or chemical synapses. However, electrical synapses are relatively uncommon in the brain, it involves neural transmission by gap junction protein enabling cytoplasmic connectivity between pre-and post-synaptic neurons, mode of neuron transmission by electrical synapses is rapid. In contrast, the neuronal transmission via chemical synapses is mediated by small molecules and/or neurotransmitters stored in a synaptic vesicle in the presynaptic neurons. Upon depolarization or propagation of action potential, synaptic vesicle in the presynaptic neurons releases neurotransmitter by exocytosis into the synaptic cleft where they bind the receptor in the pre-synaptic cell (autoreceptors) to regulate the synaptic release of the neurotransmitter and the post-synaptic membrane receptors. There are substantial evidences which demonstrate that structural changes coordinated with functional changes induces plasticity, in which LTP and LTD are reciprocally modified by spine density, morphology and activity of neurotransmitter. The strength of neural plasticity results from the efficiency of associative learning and memory which could be regulated by an interactive balance between excitatory and inhibitory neurons. This balance of neurotransmitter is essential for programming the brain codes to manage the well-organized cognitions (Ascoli et al., 2008). However, the details regarding the recruitment of associative memory cells including glutamatergic and GABAergic neurons (Yan et al., 2016) and the refinement of coordinate storage and retrieval of the associated memory signals is still remains to be addressed in mammalians (Bienvenu et al., 2012; Letzkus et al., 2012). Some reports suggest that in the somatosensory cortices, during associative memory, the excitatory synaptic transmission on glutamatergic neurons and their excitability are upregulated as well as that the excitability of GABAergic neurons and their input and output synaptic transmission are downregulated (Yan et al., 2016). The inhibitory synapses have been known to undergo several forms of plasticity, which provides additional source of versatility to the regulation of the neuronal network and the emergence of complex brain states. Further, inhibitory interneurons constitute about 20% of cortical neurons and they secrete the neurotransmitter GABA (Markram *et al.*, 2004). The GABAergic inhibitory interneurons are made of multiple varied populations, three primary populations: the Ca^{2+}-binding protein parvalbumin (PV), neuropeptidesomatostatin, and the ionotropic serotonin receptor. Among a network of excitatory neurons, GABA regulates neuronal excitability (Swadlow 2003), integration (Pouille and Scanziani 2001), generate temporal synchrony, and oscillation (Somogyi and Klausberger 2005). Also during development, these interneurons regulate cell migration via GABA (Luhmann *et al.*, 2015), differentiation of neurons, and experience-dependent refinement of neuronal connectivity (Ben-Ari 2002a; Hensch and Fagiolini 2004). During the induction of inhibitory short- and long-term synaptic plasticity, the cellular and molecular mechanisms occurring at inhibitory synapses. At the presynaptic side, retrograde synaptic messengers modulate GABA release (Younts and Castillo, 2014), whereas in postsynaptic plasticity there are changes that occur in the number of gating properties of post-synaptic GABAA receptors (Kurotani et al., 2008; Petrini et al., 2014; Flores et al., 2015). Thus interneuron has a potential to modify synapses both terms of molecular machinery as well as structurally.

3. Activity-Dependent Synaptic Plasticity in Interneurons

In the brain, neuron communication is a chemically driven phenomenon which is controlled by number of biochemical signaling mechanism that may either excite or inhibit any particular neural pathway for an efficient behavioral response. Amongst enormous chemical characteristics, brain ability to remodel itself and adapt towards a variety of active functions forms the basis of brain capability in learning, memory, sensory-motor performance, language and speech production. There are several study which elucidate the mechanisms of activity-driven changes in synaptic strength and understand its physiological significance as the brain substrate for learning and memory (Malenka and Bear, 2004; Mayford et al., 2012). Activity-dependent synaptic plasticity has been observed in both the excitatory and inhibitory synapse. Structural changes have been observed in excitatory synapse as a result of activity. Actin-dependent enlargement of the postsynaptic density occurs in activity-driven structural changes in glutamatergic synapse (Matus, 2000; Honkura et al., 2008). *In vivo* studies have shown that in response to synaptic activity patterns induced by learning behavior, the dendritic spines, the morphological correlate of glutamatergic synapses in excitatory neurons are formed and eliminated (Xu et al., 2009) and the formation of durable memories is directly correlated with the stability as well as formation of new excitatory synapses. The structural rearrangement of glutamatergic synapses is critically involved in the brain processes leading to learning and memory formation (Caroni et al., 2012). Similar to glutamatergic neurotransmission, activity can alter the efficiency of synaptic transmission in inhibitory synapses.

Despite the potential of inhibitory neurotransmission to deeply alter the function of cortical networks, the plasticity in inhibitory circuits has received relatively less attention than its excitatory counterpart. Studies have unravelled different forms of plasticity in inhibitory γ-aminobutiric acid (GABA) releasing neurons and this has been proven to be unexpectedly successful as well as covering many different aspects of the physiological properties of inhibitory cell's, including glutamatergic inputs, dendritic and axonal structure, passive properties and GABAergic synapses onto target cells (Kullmann et al., 2012). Research carried out during the last two decades has made clear that inhibitory synapses undergo short- and long-term forms of plasticity and numerous examples of activity dependent changes in synaptic strength of GABAergic synapses have been described in different brain areas, including hippocampus and cortex (Gaiarsa et al., 2002; Castillo et al., 2011; Méndez and Bacci, 2011). The complex and varied collection of pre- and postsynaptic mechanisms that underlie the induction and expression of GABAergic synapse plasticity mirrors the heterogeneous nature of different inhibitory neuron subtypes that form cortical GABAergic synapses (Méndez and Bacci, 2011).The molecular composition and structure of GABAergic synapses show a high degree of dynamism as similar to that of excitatory synapses (Kittler and Moss, 2003; Lévi et al., 2008; Chen et al., 2012; van Versendaal et al., 2012). The changing nature of inhibitory synapse structure raises the possibility that functional alterations in inhibitory neurotransmission may occur through structural rearrangements. Modifications in the structure of GABAergic synapses with two major consequences: alteration of synaptic size and morphology and formation and elimination of inhibitory contacts occurs due to synaptic activity driven functional changes of inhibitory neurotransmission. Studies have shown that both *in vitro* (Lushnikova et. al., 2011) and *in vivo* (Nusser et al., 1998), the rise in synaptic activity levels increased inhibitory synaptic junctional area, complexity and proportion of somatic cell surface covered with inhibitory postsynaptic density (PSD). Activity leads to alteration in the synaptic ultrastructure in GABAergic interneurons (Flores and Méndez 2014) Thus there is a direct positive relation exists between synaptic activity and GABA synapse formation in the adult somatosensory cortex.

CONCLUSION

The mechanism underlying the role of interneurons are important in various functions and the inhibitory circuits can be modified through various environmental stimuli. Several structural and morphological changes occur due to activity-dependent synaptic plasticity in GABAergic synapses of mammalian cortices and hippocampal. Thus role of synaptic activity and experience in the induction of structural alterations of inhibitory synapses is enormous. However, various molecular mechanisms underlie this form of plasticity should be assessed in further detail to elucidate the role of different inhibitory neuron subtypes that configure cortical networks and potential physiological consequences of the structural remodeling of GABAergic synapses.

REFERENCES

Ascoli G. A., Alonso-Nanclares L., Anderson S. A., Barrionuevo G., Benavides-Piccione R., Burkhalter A., Buzsáki G., Cauli B., Defelipe J., Fairén A., Feldmeyer D., Fishell G., Fregnac Y., Freund T. F., Gardner D., Gardner E. P., Goldberg J. H., Helmstaedter M., Hestrin S., Karube F., Kisvárday Z. F., Lambolez B., Lewis D. A., Marin O., Markram H., Muñoz A., Packer A., Petersen C. C., Rockland K. S., Rossier J., Rudy B., Somogyi P., Staiger J. F., Tamas G., Thomson A. M., Toledo-Rodriguez M., Wang Y., West D. C., Yuste R. (2008) Petilla terminology: nomenclature of features of GABAergic interneurons of the cerebral cortex. *Nat Rev Neurosci* 9: 557-568.

Bear M. F., Connors B. W., Paradisio M. A. (2007) *Neuroscience: Exploring the Brain*, 3rd ed. Lippincott, Williams & Wilkins.

Ben-Ari Y. (2002), Excitatory actions of gaba during development: the nature of the nurture. *Nat Rev Neurosci* 37: 728-739.

Bienvenu T. C., Busti D., Magill P. J., Ferraguti F., Capogna M. (2012) Cell-type-specific recruitment of amygdala interneurons to hippocampal theta rhythm and noxious stimuli *in vivo*. *Neuron* 74: 1059-1074.

Caroni P., Donato F., Muller D. (2012) Structural plasticity upon learning: regulation and functions. *Nat Rev Neurosci* 13, 478-490.

Castillo P. E., Chiu C. Q., Carroll R. C. (2011) Long-term plasticity at inhibitory synapses. *Curr Opin Neurobiol* 21: 328-338.

Chen J. L., Villa K. L., Cha J. W., So P. T., Kubota Y., Nedivi E. (2012) Clustered dynamics of inhibitory synapses and dendritic spines in the adult neocortex. *Neuron* 74: 361-373.

Flores C. E., Nikonenko I., Mendez P., Fritschy J. M., Tyagarajan S. K., Muller D. (2015) Activity-dependent inhibitory synapse remodeling through gephyrin phosphorylation. *Proc Natl Acad Sci USA* 112: E65-E72.

Gaiarsa J. L., Caillard O., Ben-Ari, Y. (2002) Long-termplasticity at GABAer-gic and glycinergic synapses: mechanisms and functional significance. *Trends Neurosci* 25: 564-570.

Hensch T. K., Fagiolini M. (2004) Excitatory-inhibitory balance andcritical period plasticity in developing visual cortex. *Prog Brain Res* 147: 115-124.

Honkura N., Matsuzaki M., Noguchi J., Ellis-Davies G. C., Kasai H. (2008), The subspine organization of actin fibers regulates the structure and plasticity of dendritic spines. *Neuron* 57, 719-729.

Horlemann J., Zieglgänsberger W. (2009) Pain processing, pain memory and quality of sleep. *Dtsch Med Wochenschr (German Medical Weekly)* 134 (supplement 4):S127-S131.

Kurotani T., Yamada K., Yoshimura Y., Crair M. C., Komatsu Y. (2008) State-dependent bidirectional modification of somatic inhibition in neocortical pyramidal cells. *Neuron* 57: 905-916.

Kittler J. T., Moss S. J. (2003). Modulation of GABAA receptor activity by phosphorylation and receptor trafficking: implications for the efficacy of synaptic inhibition. *Curr Opin Neurobiol* 13: 341-347.

Kullmann D. M., Moreau A. W., Bakiri Y., Nicholson E. (2012). Plasticity of inhibition. *Neuron* 75: 951-962.

Letzkus J. J., Wolff S. B., Meyer E. M., Tovote P., Courtin J., Herry C., Lüthi A. (2012). A disinhibitory microcircuit for associative fear learning in the auditory cortex. *Nature* 480: 331-335.

Lévi S., Schweizer C., Bannai H., Pascual O., Charrier C., Triller A. (2008) Homeostatic regulation of synaptic GlyR numbers driven by lateral diffusion. *Neuron* 59: 261-273.

Luhmann H. J., Fukuda A., Kilb W. (2015) Control of cortical neuronal migration by glutamate and GABA. *Front Cell Neurosci.* 9:4.

Lushnikova I., Skibo G., Muller D., Nikonenko I. (2011) Excitatorysynaptic activity is associated with a rapid structural plasticity of inhibitory synapses on hippocampal CA1 pyramidal cells. *Neuropharmacol* 60: 757-764.

Malenka R. C., Bear M. F. (2004) LTP and LTD: an embarrassment of riches. *Neuron* 44: 5-21.

Markram H., Toledo-Rodriguez M., Wang Y. Gupta A. (2004) Interneurons of the neocortical inhibitory system. *Nat Rev Neurosci* 5: 793-807.

Matus A. (2000) Actin-based plasticity in dendritic spines. *Science* 290: 754-758.

Mayford M., Siegelbaum S. A., Kandel E. R. (2012) Synapses and memory storage. *Cold Spring Harb Perspect Biol* 4: a005751.

Méndez P., Bacci A. (2011), Assortment of GABAergic plasticity in the cortical interneuron melting pot. *Neural Plast*. 2011: 976856.

Nusser Z., Hájos N., Somogyi P., Mody I. (1998) Increased number of synaptic GABA(A) receptors underlies potentiation at hippocampal inhibitory synapses. *Nature* 395: 172-177.

Petrini E. M., Ravasenga T., Hausrat T. J., Iurilli G., Olcese U., Racine V., Sibarita J. B., Jacob T. C., Moss S. J., Benfenati F., Medini P., Kneussel M., Barberis A. (2014) Synaptic recruitment of gephyrin regulates surface GABAA receptor dynamics for the expression of inhibitory LTP. *Nat Commun* 5: 3921.

Pouille F., Scanziani M. (2001) Enforcement of temporal fidelity in pyramidal cells by somatic feed-forward inhibition. *Science* 293: 1159-1163.

Riedel G., Platt B., Micheau J. (2003) Glutamate receptor function in learning and memory. *Behav Brain Res*. 140: 1-47.

Somogyi P., Klausberger T. (2005) Defined types of cortical interneurone structure space and spike timing in the hippocampus. *J Physiol* 562: 9-26.

Swadlow H. A. (2003) Fast-spike interneurons and feed forward inhibition in awake sensory neocortex. *Cereb Cortex* 13: 25-32.

Van Versendaal D., Rajendran R., Saiepour M. H., Klooster J., Smit-Rigter L., Sommeijer J. P., De Zeeuw C. I., Hofer S. B., Heimel J. A., Levelt C. N. (2012) Elimination of inhibitory synapses is a major component of adult ocular dominance plasticity. *Neuron* 74: 374-383.

Xu T., Yu X., Perlik A. J., Tobin W. F., Zweig J. A., Tennant K., Jones T., Zuo Y. (2009) Rapid formation and selective stabilization of synapses for enduring motor memories. *Nature* 462: 915-919.

Yan F., Gao Z., Chen P., Huang L., Wang D., Chen N., Wu R., Feng J., Cui S., Lu W., Wang J. H. (2016) Coordinated plasticity between barrel cortical glutamatergic and GABAergic neurons during associative memory. *Neural Plast* 2016: 1-20.

Younts T. J., Castillo P. E. (2014) Endogenous cannabinoid signaling at inhibitory interneurons. *Curr Opin Neurobiol* 26: 42-50.

In: Synaptic Plasticity
Editor: Sraboni Chaudhury
ISBN: 978-1-53613-606-7

Chapter 2

An Integrated Mathematical Model of Synaptic Plasticity Induction by Postsynaptic Ca^{2+} Influx through the Interactions of AMPA and NMDA Receptors

Florent Petitprez[1,2], Don Kulasiri[1,*] and Jingyi Liang[1]
[1]Centre for Advanced Computational Solutions (C-fACS),
Lincoln University, Christchurch, New Zealand
[2]Ecole polytechnique, Palaiseau, France

Abstract

N-methyl-D-aspartate receptors (NMDAR) dependent Long-Term Potentiation and Depression (LTP and LTD) are widely studied forms of synaptic plasticity and are believed to be crucial mechanisms related to memory formation. In CA1 of the hippocampus, Ca^{2+} influx in the postsynaptic spine, triggered by glutamate stimulation of the synapse

* Corresponding Author: Don.Kulasiri@lincoln.ac.nz.

through a complex process involving NMDARs and α-amino-3-hydroxy-5-methyl-4-isoxazolepropionic acid receptors (AMPARs), is a key component of both LTP and LTD inductions. Postsynaptic Ca^{2+} activates calmodulin (CaM), which triggers a signalling pathway leading to LTD or LTP if the synapse is stimulated by the low- or high-frequency signals, respectively. In this study, we develop an integrated mathematical model, using the frequency of the glutamate stimulation as the input, to explain the Ca^{2+} dynamics as a result of the interaction of AMPAR and NMDAR and the subsequent Ca^{2+}-dependent activation of CaM. We discuss the development of the model and the linkage between a few model components, including receptor dynamics, Ca^{2+} dynamics, and CaM activation, through an electrophysical model of the postsynaptic membrane. Using this model, we investigate the receptor behaviours and Ca^{2+} patterns following low- and high- frequency stimulations. The levels of the CaM activation as the result of the distinct Ca^{2+} patterns may explain the frequency dependence of LTD/LTP inductions.

Keywords: calmodulin, long-term potentiation, long-term depression, AMPAR NMDAR

1. INTRODUCTION

The nervous system is composed of neurons, which communicate with each other through synapses, and they can be thought of as one-way information channels. All synapses through the nervous system have different strengths and the strength of a synapse is shown by the amplitude of the response to a specific input signal. For example, the synaptic strength can be measured by recording the amplitude of the Excitatory Post-Synaptic Current (EPSC), the current through the membrane of the post-synaptic neuron after a stimulation. This synaptic strength changes with time, therefore dynamic in character (Citri and Malenka, 2007). The process by which the synaptic strength changes is known as synaptic plasticity and it is caused by the changes in the biochemical properties, such as the number of receptors (Citri and Malenka,2007(Citri and Malenka,2007; Mayford et al., 2012), the size of the synapse (He et al., 2014) and the receptor conductance ￼Mayford et al., 2012; Derkach et al., 2007; Lisman et al., 2012).

Several studies have linked the synaptic plasticity to memory formation (Mayford et al., 2012; Bear et al., 2006; Martin et al., 2000), and therefore it is of particular importance to study the mechanisms related to synaptic plasticity (Citri and Malenka, 2007). Several types of mechanisms are identified, depending on their duration and whether they increase or decrease the synaptic strength. Synaptic plasticity is bidirectional, the high frequency stimulation increases the synaptic strength, or potentiation, but the low frequency stimulation decreases the synaptic strength, or depression (Citri and Malenka, 2007(Citri and Malenka, 2007).

Long-Term Potentiation (LTP) and Long Term Depression (LTD) are particularly important long-lasting phenomena: LTP is characterized experimentally by an increase in the amplitude of the response, and by a diminution of the response latency, upon the application of a repetitive short pulse stimulation protocol or a high frequency stimulation (usually, it is experimentally induced using a 100 Hz signal); and the opposite changes are triggered by a prolonged low-frequency (under 3 Hz) to form LTD (Bear and Abraham, 1996; Dudek and Bear, 1992; Mulkey and Malenka 1992). LTP has a strong relationship with memory formation: it has been shown that both LTP and memory formation lead to the creation of new synapses (Engert and Bonhoeffer, 1999; Morgado-Bernal, 2011) and the blocking of LTP decreases memory formation in mammal (Morris, 1989; Abel et al., 1997; Chang et al., 1999). Several studies have also shown that LTP is impaired in the patients with Alzheimer's disease. The experiments described in (Nalbantoglu et al., 1997) shows that transgenic mice expressing the human C-terminus of the Alzheimer amyloid precursor protein have lower LTP amplitudes and persistency than those exhibited by control mice. The study described in (Chen et al., 2000) demonstrates the inhibited LTP induction in rat hippocampal slices after perfusion of amyloid beta-peptides. Hence it is crucial to understand the biomolecular processes that produce and sustain LTP in order to elucidate the mechanisms of memory formation. Hippocampal synaptic plasticity is composed of two phases: early phase occurs within seconds after the triggering signal and it activates signalling pathways which affects the post-synaptic receptors (Lisman et al., 2012). Late phase follows the early phase and maintains its effects on a longer duration by enhancing gene expression and translation, and activating cytoskeleton proteins to remodel the synaptic architecture (Bliss and Collingridge, 1993; Frey et al., 1993). LTP and LTD induction mechanisms in glutamergic hippocampal synapses have been widely studied [1]. The activation of N-methyl-D-aspartate receptors (NMDARs) enable a Ca^{2+} ions influx in the postsynaptic density (PSD) of the post-synaptic spine. Once the Ca^{2+} influx in the spine is triggered, Ca^{2+} ions bind rapidly to calmodulin (CaM), a protein acting as a Ca^{2+} detector. The newly formed CaM bound with all four Ca^{2+} sites (Ca^{2+}/CaM complex) activates a number of downstream targets. The frequency of the signals selects the downstream targets, kinases or phosphatases, leading to the LTP or LTD inductions, respectively (He et al., 2014). NMDARs are sometimes called "coincidence detectors," as their opening is both agonist and voltage-dependent. Their voltage-dependent block by magnesium Mg^{2+} ions is removed as a result of the activation of α-amino-3-hydroxy-5-methyl-4-isoxazolepropionic acid receptors (AMPARs) (Cull-Candy, 2007), which depolarises the post-synaptic membrane to unblock NMDARs. Several studies have modelled the single-channel behaviour of NMDARs or AMPARs (Robert and Howe, 2003; Schorge et al., 2005) or CaM and Ca^{2+}/calmodulin-dependent protein kinase II (CaMKII) activation (Holmes 2000; Chiba et al., 2008). These models are based on the biochemical functioning of the receptors and proteins, and relate their conformation changes by a set of states linked by biochemical reactions. To have an overall view of the biochemical mechanisms underlying LTP and LTD, however, we need a model accounting not only for a single type of receptor or protein, but a complete integrated model highlighting the interactions between the components responsible for the induction of early phase synaptic plasticity in hippocampal glutamergic synapses. In this paper, we develop such an integrated model of NMDARs, AMPARs, Ca^{2+} dynamics and subsequent binding to CaM molecules in PSD, in response to various glutamate stimulations, to understand the Ca^{2+} dynamics in PSD in response to the different signals that activate synaptic plasticity. The biochemical and electrophysiological model thus developed allows us to study quantitatively the Ca^{2+} dynamics in PSD, without imposing any particular membrane potential or current, as both values are interdependently evolving in our model, as they do *in vivo*. In particular, we study the influence of a range of stimulation frequency on the Ca^{2+} dynamics in PSD. Our results reveal a break in the pattern followed by the kinetics of Ca^{2+} in PSD between low frequencies under 3 hertz, and high frequencies over 5 hertz indicating that LTP induction requires a high Ca^{2+} amplitude which is accumulated through the multiple closely applied stimuli. As a result, the Ca^{2+}/CaM complex is only significantly formed in response to high frequency stimulation.

2. METHODS

Several studies have modelled the behaviour of the receptors or proteins involved in E-LTP induction. We integrate the existing and validated models of the parts of the system through the electrophysiological modelling. Our model contains two compartments describing the synaptic cleft and the post-synaptic spine. The synaptic cleft compartment, of which the volume was fixed to 0.004 μm^3, contains magnesium and Ca^{2+} ions and neurotransmitters glycine and glutamate. The spine has a volume fixed to 0.05 μm^3 and contains the internal Ca^{2+} ions and CaM. The PSD volume is about 10% of the spine volume, contains the AMPARs and NMDARs. The average numbers/concentrations of every components across a sufficiently large number of neuronal clefts and spines are assumed in this study (Table 1) and the assumed values are taken from (Chiba et al., 2008) and (Sheng and Hoogenraad 2007).

In the following contents of this section, we first describe the mathematical methods used to develop the model and then we explain the essential components, as well as their associated processes leading to synaptic plasticity, involved in the model.

Table 1. Numbers of molecules of the different biochemical components of the system

Parameter	Value	Reference
Number of AMPARs	100	(Sheng & Hoogenraad, 2007)
Number of NMDARs	20	(Sheng & Hoogenraad, 2007)
Concentration of CaM	6 uM	(Chiba et al., 2008)

2.1. Mathematical Methods

2.1.1. Mass Action Rate Law

Mass action rate law describes the reaction rate of an elementary reaction, which involves a single mechanistic step/state transition to form the product. Mass action rate law generally states that the rate of a biochemical reaction is proportional to the quantity of the participating reactant (Waage and Gulberg). A reversible elementary reaction is given in the following reaction scheme:

$$\alpha A + \beta B \underset{k_{off}}{\overset{k_{on}}{\rightleftarrows}} \sigma S + \tau T, \tag{1}$$

where A and B are the reactants or products, S and T are the products or the reactants for the forward or backward reactions, respectively. α, β, σ and τ are the numbers of corresponding molecules involved in the reaction, and *ks* are the reaction rate constants indicating the pace of the corresponding reactions. The corresponding reaction rates (V_f or V_b) are given by Eqs. (2 and 3):

$$V_f = k_{on}[A]^{\alpha}[B]^{\beta}, \tag{2}$$

$$V_b = k_{off}[S]^{\sigma}[T]^{\tau}. \tag{3}$$

Therefore, the dynamical change of A, B, S and T (concentrations of species are denoted by [A], [B], [S] and [T]) are represented by a set of ordinary differential equations (ODEs) as given by Eqs. (4-7):

$$\frac{d[A]}{dt} = -\alpha\left(V_f - V_b\right), \tag{4}$$

$$\frac{d[B]}{dt} = -\beta\left(V_f - V_b\right), \tag{5}$$

$$\frac{d[S]}{dt} = \sigma\left(V_f - V_b\right), \text{ and} \tag{6}$$

$$\frac{d[T]}{dt} = \tau\left(V_f - V_b\right). \tag{7}$$

2.1.2. Electrophysical Modelling

An electrophysical model of the post-synaptic membrane (Figure 1) is developed based on the work by Hodgkin and Huxley (Hodgkin and Huxley, 1952) to combine the two types of channel receptor (NMDARs and AMPARs) so that the membrane potential can be linked to the EPSC. EPSC is defined as the sum of the currents through NMDARs and AMPARs. The current through the AMPAR is calculated with the following equation:

$$i_{AMPAR} = \sum_{i=2}^{4} n_{Oi} \cdot g_{Oi} \cdot \left(V_m - V_{rev}^{AMPAR}\right) \tag{8}$$

where *Oi* is an opening state of AMPARs that allows ions to pass through the membrane and n_{Oi} is the number of the corresponding opening AMPARs. According to the work by Robert and Howe, 2003, there are three opening states for AMPARs, O2, O3 and O4, corresponding to 2, 3, and 4 glutamate molecules that are bound, each has the single channel conductance (g_{Oi}) of 9*pS*, 15*pS* and 21*pS*, respectively. V_m is the membrane potential with a rest level of $65mV$ and V_{rev}^{AMPAR} is the AMPAR reversal potential, set at $0mV$ (Allam et al., 2012). Current through the NMDARs was calculated in the same way:

$$i_{NMDAR} = \sum_{i=1}^{2} n_{Openi} \cdot g_{Openi} \cdot (V_m - V_{rev}^{NMDAR}) \tag{9}$$

where n_{Openi} is the number of NMDARs in the state Openi, g_{Openi} is the single channel conductance set at $40pS$ and $247pS$ for two opening states of NMDAR, Open1 and Open2, respectively (Schorge et al., 2005; Ambert et al., 2010). V_{rev}^{NMDAR} is the NMDAR reversal potential, set at $0.7mV$ (Allam et al., 2012). However, different from the AMPAR states defined, Open1 and Open2 states of NMDARs bind to the same number of agonists, which are glutamate and glycine (Schorge et al., 2005). The dynamical transitions among the states of receptors will be discussed in details in the next section.

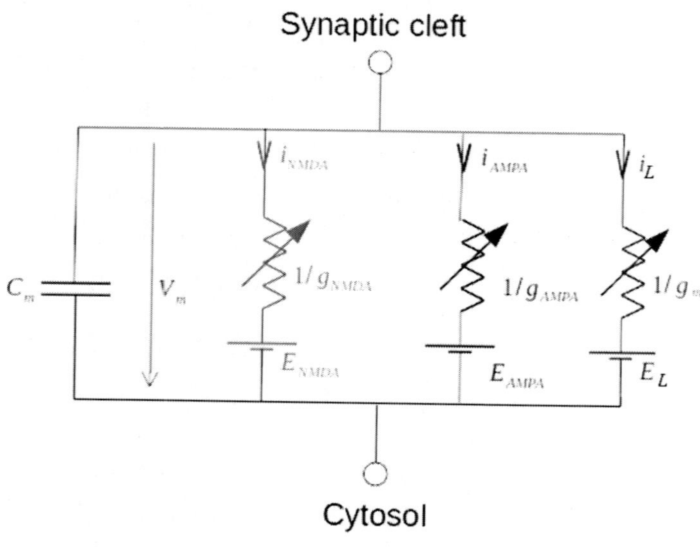

Figure 1. Electrophysical model of the post-synaptic membrane. The green branch represents NMDARs and the red one represents AMPARs. The variable resistances, characterized by their conductance, are dependent on the number of open receptors as well as the open state they are in. The left branch represents the capacitance of the post-synaptic membrane.

By assuming the total current flowing through the membrane is zero at rest, the following relationship between the membrane potential and the EPSC can be derived:

$$C_m \frac{dV_m}{dt} = -(EPSC + i_{leak}) = -(i_{AMPAR} + i_{NMDAR} + i_{leak}), \qquad (10)$$

where C_m is the membrane capacitance, calculated using the specific membrane capacitance of 1uF/cm^2 (Huguenard and McCormick, 1994), and i_{leak} is the leak current.

2.2. Mathematical Model

2.2.1. Synaptic Cleft - AMPAR and NMDAR Models

We modelled AMPARs based on mass action rate law according to the kinetic model developed by (Robert and Howe, 2003), who acknowledges the four subunits composition of the AMPAR (Gouaux 2004). The state of the receptor is defined according to the number of glutamate molecules that are bound, whether the receptor is closed or open, and the number of desensitized subunits (Figure 2A). We modelled NMDAR based on mass action rate law according to the kinetic model developed by (Schorge et al., 2005). This model counts the number and the nature of ligands that are bound by the receptor. Once the ligands are fully bound to a NMDAR, the NMDAR undergoes transitions among desensitized and open states (Figure 2B). Further, we include the block by magnesium ions by adding two states in transition with the open states, access to which was both magnesium and voltage-dependent (Ambert et al., 2010). We modelled this by finding the kinetic parameters that change with the membrane potential. At the end, we link the AMPAR and NMDAR models by the electrophysical model. The current through the NMDARs is used to calculate the amount of Ca^{2+} ions get across the membrane in the next section. AMPARs are ligand-gated ion channels, which means they open a membrane channel that allows ions to cross it when the chemical messenger is presented and bound. In the case of AMPARs, the messenger is the glutamate. AMPARs are composed of four subunits, each of which can bind to a glutamate molecule (Allam et al., 2012). Each subunit can be desensitized, leading to the inability for the channel to open.

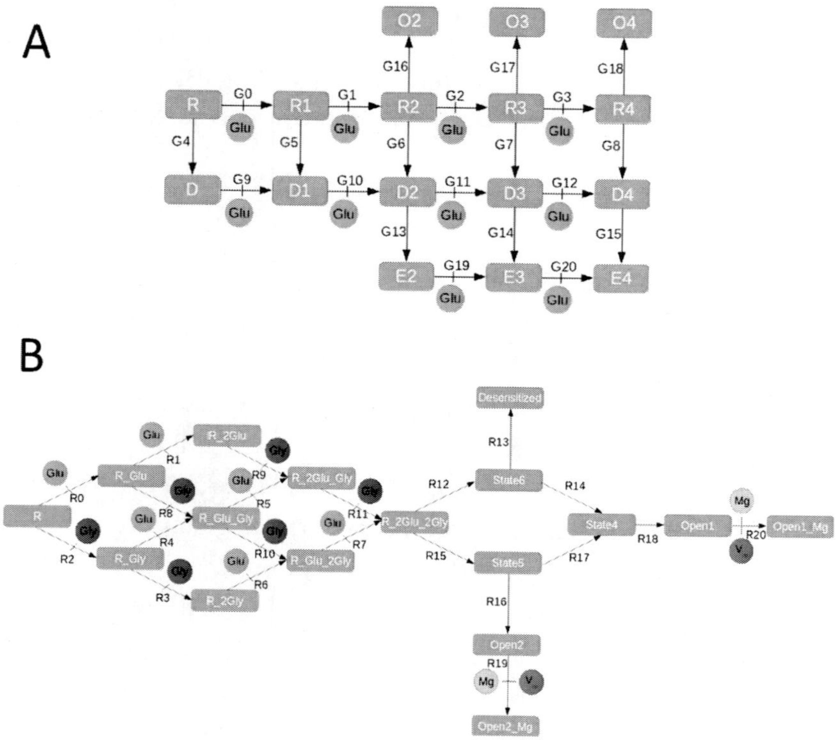

Figure 2. Structure of the AMPAR and NMDAR models. (A): Structure of the AMPAR model. States labelled R to R4 represent the receptor that is neither open nor desensitized. States O2 to O4 are the open states and states D to D4 and E2 to E4 are desensitized on one or two subunits. Contribution of glutamate (Glu) to reactions is indicated by the green circles. For more details on the different states, see appendix A. (B): Structure of the NMDAR model. States with a label starting by R are the states where the receptor is neither open nor desensitized. States 4 to 6 are transitional states. Open1 and Open2 represent the open receptor without Magnesium blockade. Open1_Mg and Open2_Mg are magnesium-blocked states and Desensitized represent the state where the receptor is desensitized. Contribution of glutamate (Glu), glycine (Gly), magnesium (Mg) and membrane potential (V_m) to reactions are respectively indicated through green, red, cyan and magenta circles.

We used the model developed by (Robert and Howe 2003) to represent this structure. The model is summarized in Figure 2A. This model considers the number of glutamate molecules bound to the receptor, with the states R (unbound) to R4 (all four subunits are bound). When 2, 3 or 4 subunits are bound to a glutamate, the receptor is able to open, which is respectively represented by the states O2, O3 and O4. Though these three states represent an open receptor, their single-channel conductance are different, hence the need of three distinct states. Here, we assume that the whole receptor is desensitized when any one of its subunits is desensitized, regardless of the number of glutamate molecules bound to sensitized or desensitized subunits. Desensitisation of one subunit is represented by the states D (desensitized with 0 glutamate bound) to D4 (desensitized with 4 glutamate bound), where it is important to note that the receptor keeps its ability to bind glutamate though the receptor is unable to open. Desensitisation of a second subunit, following the first one, can occur that is represented by states E2 (2 desensitized subunits and 2 glutamate bound) to E4 (2 desensitized subunits with 4 glutamate bound). In theory, desensitized AMPAR can have less than 2 glutamate bound and more than 2 desensitized subunits. But these cases are rare and unstable; therefore, these states are ignored. The kinetics parameters of the model were taken from (Allam et al., 2012) and are given in Table 2.

NMDARs are ligand-gated ion channels composed of four subunits, 2 of which (NR2 type) bind glutamate while the other 2 (NR1 type) bind co-agonist glycine (Cull-Candy, 2007). Once the four subunits are bound to their agonist, the channel changes conformation and opens. NMDARs are sometimes qualified of coincidence detectors, as their ability to let ions through the membrane is both agonist and voltage dependent. Indeed, at polarized membrane states, magnesium Mg^{2+} ions are able to bind inside the channel, causing a blockade. When the post-synaptic membrane is depolarized (by the opening of AMPARs), the magnesium stops blocking the open channel and lets ions cross the membrane. Of particular interest here is the ability for Ca^{2+} ions to go through the NMDARs. We used the model developed by (Schorge et al., 2005) to compute the behaviour of NMDARs. This model is summarized in Figure 2B. The original model

(Schorge et al., 2005) is composed of 9 states, labelled R_*n*Glu_*m*Gly, where *n* and m respectively represent the number of glutamate and glycine molecules that are bound to the receptor. When all four subunits are bound to their agonist, the receptor accesses state State4 to State6, which represent conformation changes. From these states, the receptor can enter either the desensitized state, or one of the two open states that are Open1 and Open2. As for the open states of the AMPAR, they differ in their single-channel conductance.

Table 2. Kinetic parameters for the AMPAR model. Reaction labels refer to Figure 2A. All parameters are taken from [30]

Reaction	k_{On}	k_{Off}
G0	$4 \times 10mM^{-1}.ms^{-1}$	$7ms^{-1}$
G1	$3 \cdot 10mM^{-1}.ms^{-1}$	$2 \times 7ms^{-1}$
G2	$2 \times 10mM^{-1}.ms^{-1}$	$3 \times 7ms^{-1}$
G3	$10mM^{-1}.ms^{-1}$	$4 \times 7ms^{-1}$
G4	$4 \times 3.3e^{-6}ms^{-1}$	$0.001ms^{-1}$
G5	$0.017ms^{-1}$	$0.42ms^{-1}$
G6	$2 \times 0.017ms^{-1}$	$0.42ms^{-1}$
G7	$3 \times 0.017ms^{-1}$	$0.42ms^{-1}$
Reaction	k_{On}	k_{Off}
G8	$4 \times 0.017ms^{-1}$	$0.42ms^{-1}$
G9	$3 \times 10mM^{-1}.ms^{-1}$	$4.1e^{-4}ms^{-1}$
G10	$3 \times 10mM^{-1}.ms^{-1}$	$7ms^{-1}$
G11	$2 \times 10mM^{-1}.ms^{-1}$	$2 \times 7ms^{-1}$
G12	$10mM^{-1}.ms^{-1}$	$3 \times 7ms^{-1}$
G13	$0.035ms^{-1}$	$0.2ms^{-1}$
G14	$2 \times 0.035ms^{-1}$	$0.2ms^{-1}$
G15	$3 \times 0.035ms^{-1}$	$0.2ms^{-1}$
G16	$2 \times 0.55ms^{-1}$	$0.3ms^{-1}$
G17	$3 \times 0.55ms^{-1}$	$0.3ms^{-1}$
G18	$4 \times 0.55ms^{-1}$	$0.3ms^{-1}$
G19	$2 \times 10mM^{-1}.ms^{-1}$	$7ms^{-1}$
G20	$10mM^{-1}.ms^{-1}$	$2 \times 7ms^{-1}$

Table 3. Kinetic parameters for the NMDAR model. Reaction labels refer to Figure 2B. All parameters are taken from [31]

Reaction	k_{On}	k_{Off}
R0, R4, R6	$2 \times 8.3mM^{-1}.ms^{-1}$	$0.0263ms^{-1}$
R1, R5, R7	$8.3mM^{-1}.ms^{-1}$	$2 \times 0.0263ms^{-1}$
R2, R8, R9	$2 \times 10mM^{-1}.ms^{-1}$	$0.0291ms^{-1}$
R3, R10, R11	$10mM^{-1}.ms^{-1}$	$2 \times 0.0291ms^{-1}$
R12, R17	$0.0671ms^{-1}$	$0.15ms^{-1}$
R13	$0.03ms^{-1}$	$9.5e^{-4}ms^{-1}$
R14, R15	$2.03ms^{-1}$	$22.8ms^{-1}$
R16	$0.787ms^{-1}$	$11.2ms^{-1}$
R18	$35.2ms^{-1}$	$0.728ms^{-1}$
R19, R20	$\frac{426 \cdot exp(V_m/55)}{1 + 91.3 \cdot exp(V_m/21.0)}$	$61.8e^{-3} \cdot exp(-V_m/50.0) + 4280e^{-3}exp(V_m/52.7)$

To represent the blockade of the receptor by magnesium, we added states suggested in the annex of (Ambert et al., 2010). These states, Open1_Mg and Open2_Mg are linked respectively to Open1 and Open2 by kinetics that depend on both the magnesium concentration in the synaptic cleft and the membrane potential. They represent the state where the channel is open but blocked by a magnesium ion, and are therefore not considered as open states, as they do not let ions through the membrane. All kinetic parameters were taken from (Ambert et al., 2010) and are given in Table 3.

2.2.2. Post-Synaptic Spine – Ca^{2+} and CaM Models

We take consider two sources of changes onto postsynaptic intracellular Ca^{2+} level: one is the influx of Ca^{2+} ions through NMDARs and the other is the Ca^{2+} pumps that take Ca^{2+} ions out of the spine. The dynamical change of Ca^{2+} level is the difference between the inflow through NMDARs and the outflow by Ca^{2+} pumps. We model the inflow and the outflow based on the work by (Jahr and Stevens, 1993) and (Zador et al., 1990), respectively. Then, we model the binding between Ca^{2+} and CaM through four sequential steps as described in (Chiba et al., 2008).

According to the work of (Jahr and Stevens, 1993), the proportion of the current through NMDARs that is carried by Ca^{2+} ions is given by the formula:

$$\theta = \frac{1}{1+(G_0/0.5ch)} \tag{11}$$

where c is the Ca^{2+} concentration in the synaptic cleft set at 1mM, G_0 = 65.6pS and h =11.3$pS.mM^{-1}$. We can then calculate the Ca^{2+} influx through NMDARs using:

$$\left(\frac{d[Ca^{2+}]}{dt}\right)_{NMDAR} = \frac{-i_{NMDAR}\cdot\theta}{2\cdot e\cdot N_A\cdot V_{PSD}} \tag{12}$$

where $e = 1.60217657 \times 10^{-19}C$ is the elementary charge, $N_A = 6.02214129 \times 10^{23}\ mol^{-1}$ is the Avogadro constant and V_{PSD} is the volume of PSD.

The outflows through a number of Ca^{2+} pumps are modelled according to the kinetics model presented by (Zador et al., 1990). We integrated pumps obeying storable first-order Michaelis-Menten kinetics:

$$\left(\frac{\partial[Ca^{2+}]}{\partial t}\right)_{pump} = K_{max}\cdot P_s\cdot\frac{A}{V}\cdot\left(\frac{[Ca^{2+}]_{eq}}{[Ca^{2+}]_{eq}+K_d} - \frac{[Ca^{2+}]}{[Ca^{2+}]+K_d}\right) \tag{13}$$

where K_{max} = 0.2ms^{-1} is the Ca^{2+} turnover rate, P_s is the pump's surface density, A/V is the ratio of surface area to volume, K_d the dissociation constant and $[Ca^{2+}]_{eq}$ is the Ca^{2+} concentration at the equilibrium, set at 0.0001 mM. All values that are pump-specific are given in Table 4.

At every moment in the neuronal spine, there is a non-zero level of Ca^{2+}, at least 100 nM (rest). The influx of Ca^{2+} through NMDARs rise this level to as high as 100-fold. This rise in the Ca^{2+} level is detected by CaM. CaM binds up to 4 Ca^{2+} ions on 4 different sites. The CaM is fully activated when it is fully bound to Ca^{2+} ions at all four sites, forming the complex $CaMCa_4$. The affinity of CaM to Ca^{2+} ion is strong, which gives a fast binding, hence a fast activation of CaM in response to Ca^{2+} level rise. To represent CaM activation, we used the model developed by (Chiba et al., 2008), summarized in Figure 3. The order in which the CaM sites bind Ca^{2+} is always the same, as they do not have the same affinity for Ca^{2+} ions and as the binding of one ion induces conformation changes. Therefore, CaM is modelled only based on the number of ions that are bound to the protein. The kinetics of each binding are different though, as given in Table 5.

Figure 3. Structure of the CaM model. Each state represents the number of Ca^{2+} ions that are bound to CaM. Ca^{2+} contribution to the reaction is indicated by a yellow circle.

Table 4. Kinetic parameters for the Ca^{2+} pumps, referring to equation 13. The values are simplified from [35]

Pump type	Variable	Value
1	P_s	$5e^{-16}\mu mol.\mu m^{-2}$
	K_d	$0.5\mu M$
	A/V	$5.3\ \mu m^{-1}$
2	P_s	$1e^{-15}\mu mol.\mu m^{-2}$
	K_d	$20\mu M$
	A/V	$5.3\ \mu m^{-1}$

Table 5. Kinetic parameters for the CaM model. All the values are taken from [26]. The reaction labels refer to those of Figure 3

Reaction	k_{On}	k_{Off}
C1	$2.5mM^{-1}.ms^{-1}$	$0.05ms^{-1}$
C2	$88.25mM^{-1}.ms^{-1}$	$2 \times 0.05ms^{-1}$
C3	$12.5mM^{-1}.ms^{-1}$	$1.25ms^{-1}$
C4	$250mM^{-1}.ms^{-1}$	$1.25ms^{-1}$

2.2.3. Simulation and Input

To study the behaviour of the model, simulations have been run using the software COPASI (Schorge S, et al. 2005). COPASI, or COmplex PAthway SImulator is an open-source biochemical pathways simulator. It establishes the ordinary differential equations (ODEs) that regulate the time-course evolution of each element of the system. COPASI offers a range of solvers for the ODEs. We used the LSODA solver (Petzold, 1983), which alternates between stiff and non-stiff methods to deliver a numerical solution of the ODEs.

To stimulate the synapse, the ideal square signal of glutamate was approximated by the first six terms of the Fourier series to overcome numerical simulation failures internal to COPASI. In order to diminish the influence of the Gibbs phenomenon, the input signal was blocked at both 0 and 1. With these assumptions, the glutamate concentration during the time of stimulation was given by the formula:

$$[Glu](t) = max\left(0, min\left(1, \frac{4}{\pi} \cdot \sum_{k=1}^{6} \frac{sin((2k+1) \cdot \pi f t)}{2k+1}\right)\right) \tag{14}$$

where $[Glu]$ represents the glutamate concentration in the synaptic cleft and f is the frequency of the stimulation signal. The signal is theoretical for our purpose of testing the frequency dependence of Ca^{2+} level.

3. RESULTS AND DISCUSSION

3.1. Basic Behaviour of the Model

To study the basic behaviour of the system, we run a simulation under a simple condition where the synaptic cleft contains 100 μM glycine and the stimulation is a single pulse of 1 ms duration, 1 mM glutamate. The resulting opening probabilities of AMPARs and NMDARs, the EPSC triggered by stimulation, the Ca^{2+} influx in the PSD and the CaM activation are shown in Figure 4. We cannot compare the current curves directly against experimental records because the experiments are carried based on current- or voltage-clamp protocols, where either the current through the membrane or the membrane potential are fixed. In our model, both quantities evolve through time and depend on each other. But if the membrane potential is held constant, equations 8 and 9 give proportionality between the open probability of the receptors and the current. Therefore, we can compare the shape of the curve for theoretical open probability and experimental current through the membrane in voltage-clamp experiments. The kinetics of the opening of AMPARs are in good agreement with the experimental data presented by (Lester et al., 1990): We can indeed compare the shape of the theoretical open probability curve (Figure 4A) to the experimental current in a voltage-clamp experiment (Figure 1 of Lester et al., 1990). Moreover, this simulation shows that a small increase in Ca^{2+} level is unable to fully activate CaM to trigger LTP in the synapse. Although CaM rapidly binds to Ca^{2+} and transits to other bound states, the maximum level of fully active CaM ($CaMCa_4$) does not go beyond 2.91% of the total CaM that is insufficient to activate downstream proteins that requires binding to the fully active CaM (Figure 4D).

We also simulate the behaviour of the synapse when subjected to the multiple-pulse high frequency stimulus, at 100 hertz of 1mM glutamate lasting for 0.5 second in the presence of 100 μM glycine in the synaptic cleft. Opening probabilities, EPSC, Ca^{2+} influx and CaM activation are shown in Figure 5. The behaviour of NMDAR is in good agreement with experimental data (Rosenmund et al., 1995) by comparing the shape of the open probability curve of NMDARs (Figure 5A) to the experimental current (Figure 1A of Rosenmund et al., 1995). The opening probability of receptors does not show significant difference between single and multiple pulse cases (Figures 4A and 5A). Currents through NMDAR diminish over pulses, but the overall induced currents last much longer as compared to single-pulse case (Figures 4B and 5B), that result in both a greater maximum and longer duration of the resulting Ca^{2+} rise (Figures 4C and 5C). Furthermore, more CaM are bound to Ca^{2+} ions (Figures. 4D and 5D) resulting in a much higher CaM activation, with up to 56.7% are fully activated (Figure 5D).

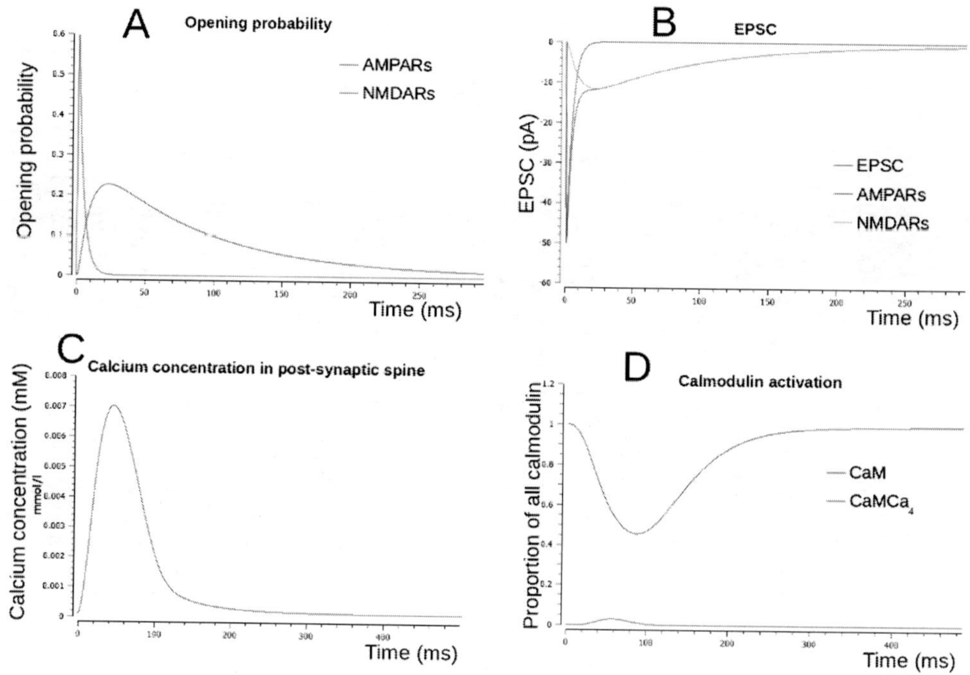

Figure 4: Results of the stimulation by a single pulse of 1 mM glutamate lasting 1 ms. (A): Open probability of AMPARs (red) and NMDARs (blue). The open probability is calculated as the ratio of the number of receptors in any open states over the total number of receptors. (B): Excitatory Post-Synaptic Current (EPSC) triggered by the pulse. The EPSC is drawn in red. The components through AMPARs and NMDARs are respectively shown in blue and green. AMPARs carry the fast response, whereas NMDARs are responsible the late one. (C): Time-course evolution of the Ca2+ concentration in the PSD. (D): Evolution of the proportion of CaM in each model state. The activated form of CaM, CaMCa4, is shown in red and remains under 0.0291 for this simulation.

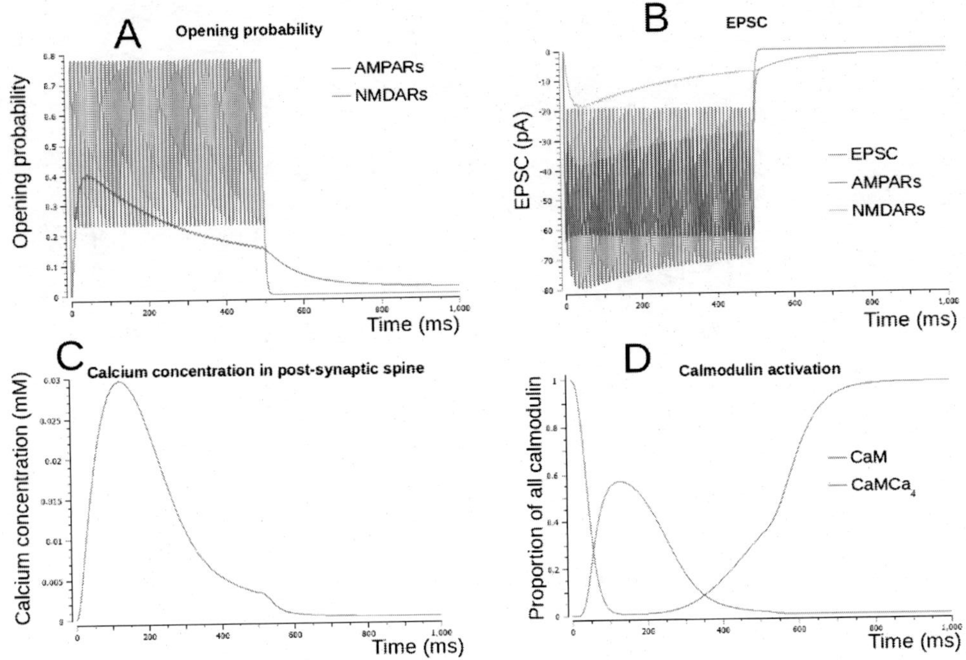

Figure 5. Results of the stimulation by a 100 Hz signal of amplitude 1 mM glutamate lasting 500 ms. (A): Open probability of AMPARs (red) and NMDARs (blue). The open probability is calculated as explained in Figure 4. (B): Excitatory Post-Synaptic Current (EPSC) triggered by the pulse. The EPSC is drawn in red. The components through AMPARs and NMDARs are respectively shown in blue and green. (C): Time-course evolution of the Ca2+ concentration in PSD. The Ca2+ concentration decreases towards its rest level before the end of the stimulation. (D): Evolution of the proportion of CaM in each model state. The activated form of CaM, CaMCa4, is shown in red and reaches 0.567 in this simulation.

3.2. Influence of the Stimulation Frequency on Ca2+ Dynamics

We use our model to study the influence of frequency on the Ca^{2+} dynamics in the PSD. We stimulate the system with 100 pulses at a range of frequencies, at 1, 2, 3, 4, 5, 10, 20, 50, 70 and 100 hertz, such that the total duration of the stimulation is dependent on the frequency chosen. We then look at two quantities: (1) the time, from the beginning of the stimulation, at which the maximum of Ca^{2+}concentration in the spine is reached, and (2) the exact value of the maximum Ca^{2+} concentration. This process is repeated with different values of glycine concentration in the synaptic cleft. The results of these computational experiments are given in Figure 6. In Figure 6, there is a clear break between frequencies under 3 Hz, and frequencies over 20 Hz for the time to reach maximum. Under 3 Hz, the maximum is reached around 140 ms with the little variation. When the frequency is over 20 Hz, the maximum is reached at a time that decreases with the frequency logarithm. This break between 10 Hz is also seen on the height of the peak of the Ca^{2+} level rise and is consistent across different glycine concentrations. Under 3 Hz, it is not significantly affected by a change in the frequency. Over 10 Hz, the height of the Ca^{2+} peak is increasing with the stimulation frequency. The behaviour of the system that enables frequency to influence the Ca^{2+} dynamics seems to act like a paired-pulse facilitation. When the frequency is low, the Ca^{2+} peak is reached around 120 ms, as shown in Figure 7. Under 3 Hz (inter-pulse interval is greater than 333ms), there is only one glutamate application occurring before this time and the Ca^{2+} level decreases before the second pulse is applied. When the frequency is over 5 Hz (inter-pulse interval is less than 200 ms), the second glutamate application starts before 200 ms, which is before the Ca^{2+} level decreases. This second application induces a greater Ca^{2+} level rise and therefore delays the Ca^{2+} level to reach the peak. However, when the frequency gets higher, the inter-pulse interval becomes more and more insufficient for the receptors to recovery from the desensitisation. Hence, a decrease in both the level of the peak and the time to reach the peak is obtained with increasing frequency beyond 20Hz (Figure 6). With respect to these different patterns for low and high stimulation frequencies, our model reveals a switch-like behaviour of frequency to lead the post-synaptic Ca^{2+} dynamics into one of two different patterns, one under 3 Hz, the second one over 10 Hz. The two patterns are distinguished by a change in both the time at which the maximum Ca^{2+} concentration is reached and the value of this maximum. Such a difference between two patterns for Ca^{2+} dynamics in the post-synaptic neuron may be the key to enable either LTD at low frequencies, or LTP at higher frequencies.

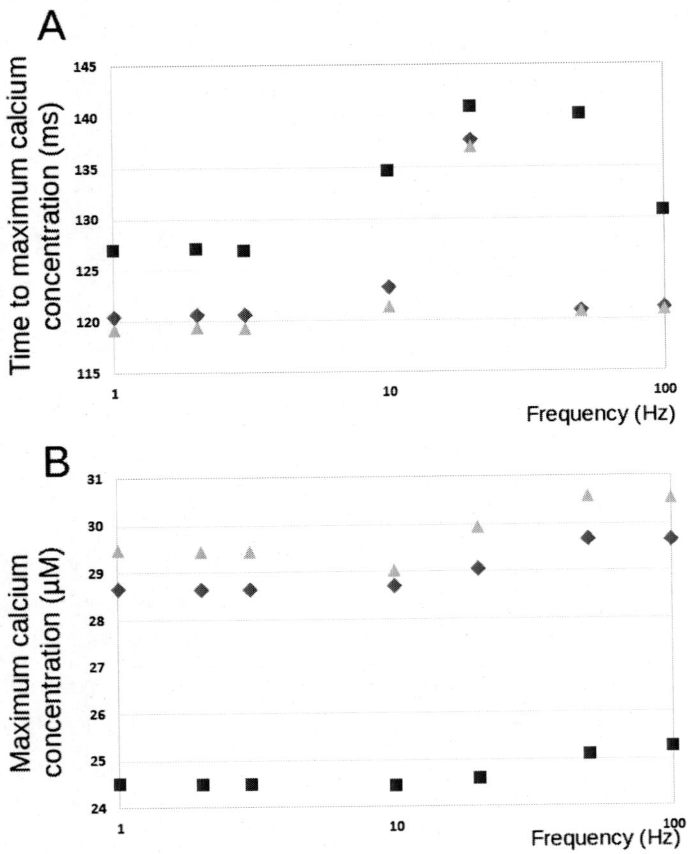

Figure 6. Evolution of Ca^{2+} dynamics following a change in the stimulation frequency for different glycine concentrations. Squares: 20 μM glycine; Diamonds: 100μM glycine; Triangles: 500 μM glycine. (A): Time to reach the maximum Ca^{2+} level after the start of the stimulation. (B): Maximum Ca^{2+} concentration in PSD.

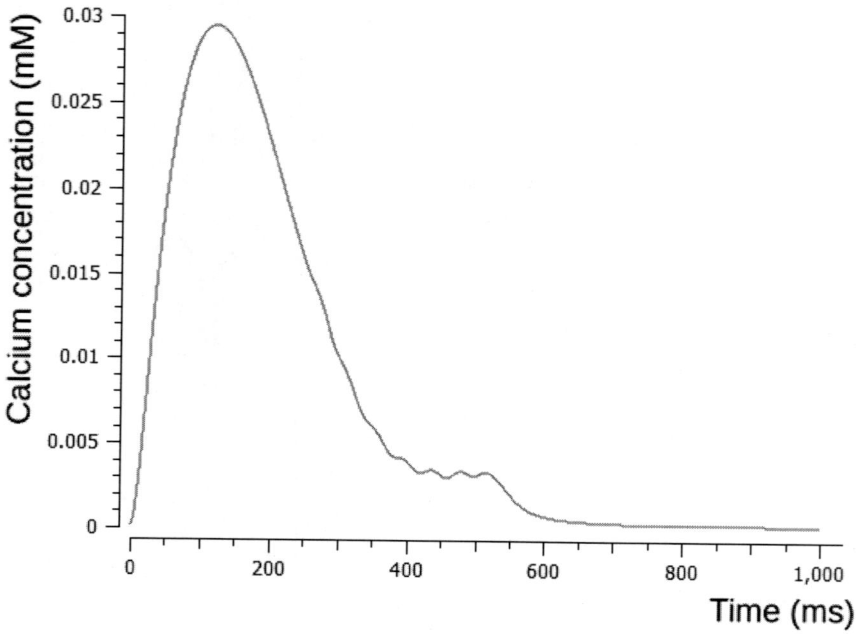

Figure 7. Ca^{2+} dynamics in PSD following a stimulation by 1mM glutamate at 2 Hz during 500 ms. The maximum Ca^{2+} concentration is reached after 119.4 ms and is at 29.45 μM.

3.3. Influence of the Stimulation Duration

In order to study whether the duration of the stimulation has a significant effect on the behaviour of our model and in particular on the Ca^{2+} dynamics, we study the impact of a change on the stimulation duration while keeping the frequency fixed. We conducted the computational experiment for 1, 5, 29 and 100 Hz and each lasts for 100, 200, 500, 1000, 2000, 5000 and 10000 milliseconds. The results are given in Figure 8. The results show that the simulation duration is not a strong factor of the Ca^{2+} dynamics as long as the simulation duration is long enough to accommodate the increase of Ca^{2+} to reach its maximum.

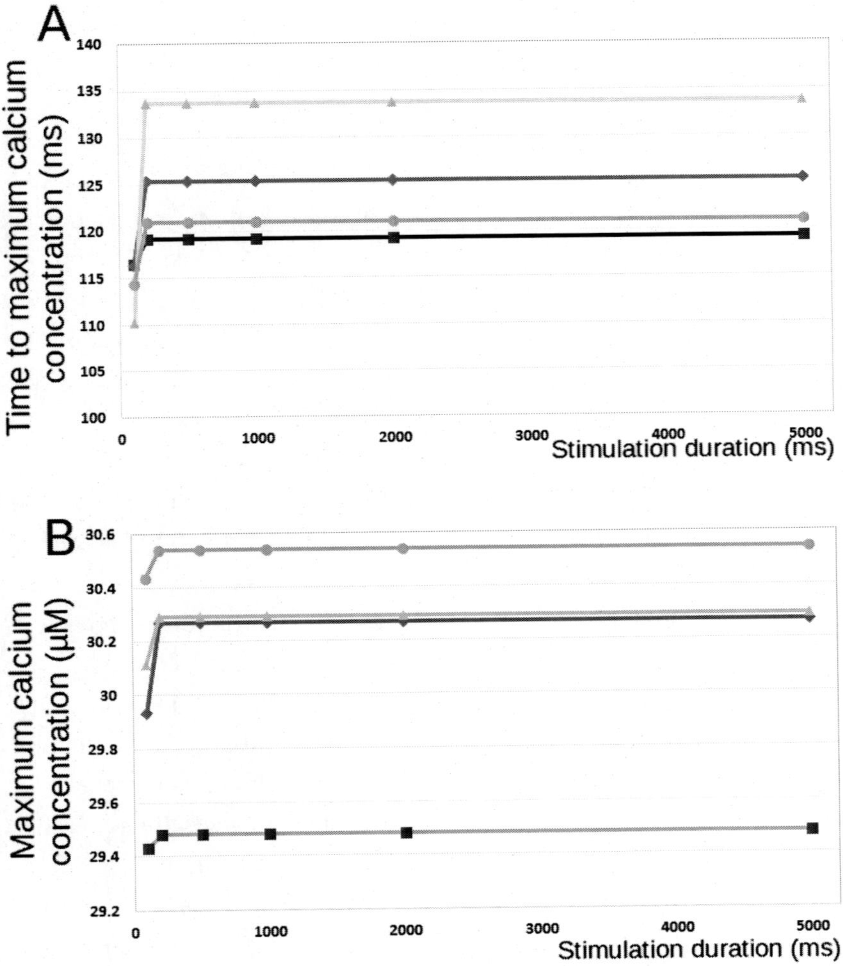

Figure 8: Effect of the stimulation duration on Ca^{2+} dynamics for different stimulation frequencies. Squares: 1Hz; Diamonds: 5 Hz; Triangles: 29 Hz; Circles: 100 Hz. (A): Time to reach the maximum Ca^{2+} level after the start of the stimulation. This value becomes a constant after the stimulation duration reaches the time to a maximum Ca^{2+} concentration of the experiment reported in figure 6. (B): Maximum Ca^{2+} concentration in PSD. It also becomes a constant at the same instant.

4. Conclusion

By developing an integrated biochemical and electrophysical model of hippocampal Ca^{2+} dynamics, we have shown that the stimulation frequency influences the Ca^{2+} dynamics in the post-synaptic neuron in a switch-like manner, and the switch-like behaviour may have a relation to induction of LTP or LTD. However, once the stimulation is longer than the time at which the maximum Ca^{2+} level is achieved, increasing the duration does not influence the Ca^{2+} peak in PSD. This absence of influence of the stimulation duration finds its origin in the electrophysical model of the post-synaptic membrane, and particularly in the branch that contains the capacitance. The capacitance allows the depolarisation of the membrane by a current influx through AMPARs and NMDARs and brings the membrane potential towards zero. There is little current flowing through the receptors at depolarized states since the current is proportional to the difference between the membrane potential and the reversal potential of the receptors and both the reversal potentials being close to zero. In particular, when the current through NMDARs is very small, there is little Ca^{2+} influx in PSD. The Ca^{2+} pumps chase out the remaining Ca^{2+} and a prolonged stimulation is unable to trigger a further entry of Ca^{2+}. To be more comprehensive regarding the LTP or LTD implications, our model needs to be further completed by adding a model of CaMKII activation, and the molecules of the signalling pathway it activates. This will enable us to conclude regarding CaM and its influence on the synapse strength. The focus of the observation may then go from Ca^{2+}, as in the model presented here to CaM or CamKII activation, which may provide further explanation of the synapse's behaviour through LTP and LTD. Although not taking into account all molecules involved in the induction of LTP and LTD, our model proves to be sufficient to explain the Ca^{2+} dynamics in PSD, and therefore the signal entry in the post-synaptic neuron. This can then be used as a basis to build further models describing what follows this Ca^{2+} entry. Indeed, the two different patterns followed by Ca^{2+} dynamics may be a key to understand the induction of the bidirectional change in synaptic strength, with one pattern leading to LTD, whereas the other would lead to LTP.

REFERENCES

Abel T, Nguyen P V, Barad M, Deuel TAS, Kandel ER, Bourtchouladze R (1997) Genetic demonstration of a role for PKA in the late phase of LTP and in hippocampus-based long-term memory. *Cell* 88: 615–626.

Allam SL, Ghaderi VS, Bouteiller JM, Legendre A, Nicolas A, Greget R, Bischoff S, Baudry M, Berger TW (2012) A computational model to investigate astrocytic glutamate uptake influence on synaptic transmission and neuronal spiking. *Front Comput Neurosci* 6: 70.

Ambert N, Greget R, Haeberle O, Bischoff S, Berger TW, Bouteiller JM, Baudry M (2010) Computational studies of NMDA receptors: differential effects of neuronal activity on efficacy of competitive and non-competitive antagonists. *Open Access Bioinformatics* 2: 113–125.

Bear MF, Abraham WC (1996) Long-Term Depression in Hippocampus. *Annu Rev Neurosci* 19: 437–462.

Bear MF, Connors BW, Paradisio MA (2006) *Neuroscience.* 3rd ed. Wolters Kluwer Health.

Bliss TVP, Collingridge GL (1993) A synaptic model of memory: long-term potentiation in the hippocampus. *Nature* 361: 31–39.

Bliss TVP, Gardner-Medwin AR (1973) Long-lasting potentiation of synaptic transmission in the dentate area of the unanaesthetized rabbit following stimulation of the perforant path. *J Physiol* 232: 357–374.

Bliss TVP, Lømo T (1973) Long-lasting potentiation of synaptic transmission in the dentate area of the anaesthetized rabbit following stimulation of the perforant path. *J Physiol* 232: 331–356.

Chang HP, Lindberg FP, Wang HL, Huang AM, Lee EHY (1999) Impaired memory retention and decreased long-term potentiation in integrin-associated protein-deficient mice. *Learn Mem* 6: 448–457.

Chen QS, Kagan BL, Hirakura Y, Xie CW (2000) Impairment of hippocampal long-term potentiation by Alzheimer amyloid beta-peptides. *J Neurosci Res* 60: 65–72.

Chiba H, Schneider NS, Matsuoka S, Noma A (2008) A Simulation Study on the Activation of Cardiac CaMKII< i> δ</i>-Isoform and Its Regulation by Phosphatases. *Biophys J* 95: 2139–2149.

Citri A, Malenka RC (2007) Synaptic plasticity: multiple forms, functions, and mechanisms. *Neuropsychopharmacology* 33: 18–41.

Cull-Candy SG (2007) NMDA Receptors. In: John Wiley & Sons, Ltd, editor. *Encyclopedia of Life Sciences.* Chichester, UK: John Wiley & Sons, Ltd.

Derkach VA, Oh MC, Guire ES, Soderling TR (2007) Regulatory mechanisms of AMPA receptors in synaptic plasticity. *Nat Rev Neurosci* 8: 101–113.

Dudek SM, Bear MF (1992) Homosynaptic long-term depression in area CA1 of hippocampus and effects of N-methyl-D-aspartate receptor blockade. *Proc Natl Acad Sci* 89: 4363–4367.

Engert F, Bonhoeffer T (1999) Dendritic spine changes associated with hippocampal long-term synaptic plasticity. *Nature* 399: 66–70.

Frey U, Huang YY, Kandel ER (1993) Effects of cAMP simulate a late stage of LTP in hippocampal CA1 neurons. *Science* 260: 1661–1664.

Gouaux E (2004) Structure and function of AMPA receptors. *J Physiol* 554: 249–253.

He Y, Kulasiri D, Samarasinghe S (2014) Systems biology of synaptic plasticity: a review on N-methyl-D-aspartate receptor mediated biochemical pathways and related mathematical models. *Biosystems* 122: 7–18.

Hodgkin AL, Huxley AF (1952) A quantitative description of membrane current and its application to conduction and excitation in nerve. *J Physiol* 117: 500–544.

Holmes WR (2000) Models of calmodulin trapping and CaM kinase II activation in a dendritic spine. *J Comput Neurosci* 8: 65–86.

Hoops S, Sahle S, Gauges R, Lee C, Pahle J, Simus N, Singhal M, Xu L, Mendes P, Kummer U (2006) COPASI—a COmplex PAthway SImulator. *Bioinformatics* 22: 3067–3074.

Huguenard J, McCormick DA (1994) *Electrophysiology of the neuron.*

Jahr CE, Stevens CF (1993) Calcium permeability of the N-methyl-D-aspartate receptor channel in hippocampal neurons in culture. *Proc Natl Acad Sci* 90: 11573–11577.

Lester RA, Clements JD, Westbrook GL, Jahr CE (1990) Channel kinetics determine the time course of NMDA receptor-mediated synaptic currents. *Nature* 346: 565–567.

Lisman JE, Yasuda R, Raghavachari S (2012) Mechanisms of CaMKII action in long-term potentiation. *Nat Rev Neurosci* 13: 169–182.

Martin SJ, Grimwood PD, Morris RGM (2000) Synaptic plasticity and memory: an evaluation of the hypothesis. *Annu Rev Neurosci* 23: 649–711.

Mayford M, Siegelbaum SA, Kandel ER (2012) Synapses and memory storage. *Cold Spring Harb Perspect Biol* 4: a005751.

Morgado-Bernal I (2011) Learning and memory consolidation: linking molecular and behavioral data. *Neuroscience* 176: 12–19.

Morris RG (1989) Synaptic plasticity and learning: selective impairment of learning rats and blockade of long-term potentiation in vivo by the N-methyl-D-aspartate receptor antagonist AP5. *J Neurosci* 9: 3040–3057.

Mulkey RM, Malenka RC (1992) Mechanisms underlying induction of homosynaptic long-term depression in area CA1 of the hippocampus. *Neuron* 9: 967–975.

Nalbantoglu J, Tirado-Santiago G, Lahsaïni A, Poirier J, Goncalves O, Verge G, Momoli F, Welner SA, Massicotte G, Julien JP, Shapiro ML (1997) Impaired learning and LTP in mice expressing the carboxy terminus of the Alzheimer amyloid precursor protein. *Nature* 387: 500–505.

Petzold L (1983) Automatic Selection of Methods for Solving Stiff and Nonstiff Systems of Ordinary Differential Equations. *SIAM J Sci Stat Comput* 4: 136–148.

Robert A, Howe JR (2003) How AMPA Receptor Desensitization Depends on Receptor Occupancy. *J Neurosci* 23: 847–858.

Rosenmund C, Feltz A, Westbrook GL (1995) Synaptic NMDA receptor channels have a low open probability. *J Neurosci* 15: 2788–2795.

Schorge S, Elenes S, Colquhoun D (2005) Maximum likelihood fitting of single channel NMDA activity with a mechanism composed of independent dimers of subunits. *J Physiol* 569: 395–418.

Sheng M, Hoogenraad CC (2007) The postsynaptic architecture of excitatory synapses: a more quantitative view. *Annu Rev Biochem* 76: 823–847.

Waage P, Gulberg CM (1986) Studies concerning affinity. *J Chem Educ. American Chemical Society* 63: 1044.

Zador A, Koch C, Brown TH (1990) Biophysical model of a Hebbian synapse. *Proc Natl Acad Sci* 87: 6718–6722.

In: Synaptic Plasticity
Editor: Sraboni Chaudhury

ISBN: 978-1-53613-606-7

Chapter 3

NEURONAL PLASTICITY AND EXPERIENCE-DEPENDENT BEHAVIORS: INSIGHTS FROM FRUIT FLIES

***Tuhin Subhra Chakraborty*[1,*], *Anindya Ganguly*[2,*] *and Abu Farhan*[3,*]**
[1]University of Michigan, Ann Arbor, US
[2]Neuroscience Research Institute
University of California, Santa Barbara, CA, US
[3]NYU Langone Medical Center, New York City, US

ABSTRACT

An organism's survival and success depends in large part on its ability to monitor changes in its environment and modify its behavior accordingly. For creatures with a nervous system, this involves transducing raw sensory information into electrical signals that are then processed internally, leading to a behavioral response. In order to adjust with the changes in the environment, neuronal populations can adjust their activity. This is known as neuronal plasticity and it leads to neuronal modulations at genetic, cellular and molecular level. At the genetic level,

---

* Authors contributed equally.

differential gene expression can alter synthesis of proteins associated with synaptic structures. Neuronal plasticity also includes changes in functional properties viz. ionic transport and electrical activity, known as functional plasticity. Structural modifications resulting in anatomical changes including circuit remodeling is also observed in some cases. However, neuronal plasticity is not restricted to higher brain functions or central nervous system, rather it also serves as a basic tool for adaptation at the levels of peripheral nervous system. The fruit fly, *Drosophila melanogaster,* provides a powerful genetic model to gain insight into the cellular and molecular basis of the development and function of nervous system. The nervous system of the fruit fly is simpler, allowing functional and genetic manipulations of individual genetically identifiable neurons. In this chapter, we review recent progress concerning neuronal plasticity in olfaction and gustation where olfactory and gustatory neuron's responses alter with internal physiological state. Olfactory learning and memory also causes synaptic plasticity dependent on long term memory (LTM) formation, whereas gustatory system shows changes in response to different nutritional status as well as mating status.

Keywords: Drosophila melanogaster, olfactory learning, long term memory

1. INTRODUCTION

Nervous system has evolved to translate stimuli into appropriate behavioral responses. To survive in an ever-changing environment, flexible adjustments in behavioral choices based on experience-dependent learning as well as internal physiological states is extremely crucial. To understand how behaviors are modulated we need to know how sensory information is represented and transferred in the brain, how it is changed by experience and physiological requirements and how the changes are reflected in motor output. One of the major challenges in neuroscience is to pin down the neuro-anatomical location where such flexibility is achieved and to understand how these changes are finally translated into behavior-modulations.

Although studies in primates and rodent brains have provided us with useful mechanistic insights in neuronal plasticity in certain brain areas, the complex organization of these circuits impedes understanding of the flow of neuronal information from sensory coding to motor output. This is where the simpler nervous system of a fruit fly comes handy, enabling a better insight into plasticity both at the level of a single neuron and an entire neuronal circuit. The neuronal circuit underlying a fruit fly's chemosensation viz. olfaction and gustation provide an ideal model system to address these questions.

In each circuit layer, group of neurons are recruited to encode stimuli that can influence the behavioral output. As in vertebrates, population coding at the sensory neuronal level in *Drosophila* is commonly involved in complex behaviors like sensory discrimination and learning. Several recent studies shed light on the circuit mechanism that underlies this plasticity. Interestingly, modulation of behavior can be invoked by plasticity at every layer, from peripheral sensory organs to the first information relay center and further on into multiple higher brain regions. Here we will focus on how behavior elicited by an external stimulus could be altered by its context. We will also highlight the neuromodulatory mechanisms by which internal physiological state regulates behavior.

2. Anatomy of Olfactory System in *Drosophila*

Antenna (third segment) and maxillary palps are the olfactory organs in flies. First, odors are transduced into electrochemical signals by an array of ~50 olfactory receptors (ORs) expressed in olfactory receptor neurons (ORNs) in the third antennal segment. ORNs are housed in porous cuticular structure called sensilla that cover the surface of the third antennal segments and the maxillary palps. Antennal sensilla are sub-divided into three morphological types-basiconics, coeloconics and trichoids-that are distributed on distinct overlapping zones (Venkatesh and Naresh Singh, 1984). Only basiconic sensilla are present on the maxillary palps (Naresh Singh and Nayak, 1985). Olfactory sensilla, which can house upto 4 ORNs, are further sub-divided into 23 functional classes- 12 antennal basiconic, 4 coeloconic, 4 trichoids and 3 maxillary palp basiconics.

Several research groups have systematically surveyed the responses of the ORNs to diverse sets of chemical stimuli. These studies reveal that depending on the response profile, ORNs can be classified as broadly tuned (responding to a vast array of odors) or narrowly tuned (activated only in the presence of a very few odors). However, most individual ORN respond to multiple ligands and most individual ligands activate multiple ORN types. Neuronal activity of the ORNs increases with increasing ligand concentration.

In general, each ORN in *Drosophila* expresses a single odorant receptor (OR) gene that specify the tuning of that neuron and a co-receptor *orco* which serves as a chaperone, guiding the cell surface expression of the functional olfactory complex and also forming the functional olfactory receptor complex along with the other OR. Recent studies have demonstrated a different class of receptors namely Ionotropic receptors (IR), that can also mediate olfactory responses to different odorant classes. Contrary to ORs, IR mediated olfactory responses are independent of orco.

Information from ORNs converge onto synaptic sub-compartments called glomeruli within the antennal lobe (AL). Axon from each ORN synapses with both olfactory projection neurons (PNs) and local interneurons (LNs) in a single glomerulus. PNs relay olfactory information gathered at the periphery to higher processing centers in the *Drosophila* protocerebrum namely mushroom body (MB) and lateral horn (LH). The MB is involved in olfactory learning and memory. The role of the LH is less clear, but it has been implicated in innate behaviors as well as multimodal integration of sensory information. The neuronal circuitry that regulate behavioral output following olfactory coding in the brain in flies is yet to be fully understood. However, it can be postulated that both the mushroom body and lateral horn output neurons (MBONs and LHONs) likely synapse with downstream neurons to activate appropriate behavioral motor programs.

3. Anatomy of Gustatory System in *Drosophila*

Drosophila bears specialized hair like structures known as taste bristles or taste sensilla in numerous parts of their body all of which could function at least to some extent as a chemosensor or a taste organ. Besides they also contain taste pegs and pharyngeal taste sensilla both of which are non-hair like sensory structures.

The main taste organs of the fly are the labellum, the pharyngeal organs and the tarsi. Other than these, they also have taste bristles on the anterior wing margins and females on their ovipositors.

Each sensillum typically houses multiple taste neurons along with a mechanosensory neuron and support cells at the base. Although the taste bristles contain up to four taste neurons, taste sensilla in the pharynx can have up to ten taste neurons. At the tip of the sensillum there is a pore through which chemicals can diffuse into the sensillary lymph present within. The dendrites of the taste neurons bear taste receptors on them, which are activated when they encounter chemical compounds. The taste neurons are classified based on the class of tastants they respond to. The main taste neuron classes are sweet-sensing, bitter-sensing, low-salt-sensing, and water-sensing. The sensitivity of a taste neuron is an outcome of the taste receptors present in them. While gustatory receptors or GRs which is structurally similar to ORs, being the first taste receptors to be identified in flies, are most extensively studied, recent studies have revealed the role of pickpocket channels (ppk), IRs and TRP (transient receptor potential) channels in conferring sensitivity to different tastant classes.

The organization of taste sensilla on the different taste organs is highly stereotypical. Axons of taste neurons from the different taste organs project to different parts of the sub-esophageal zone and/or the ventral nerve chord where they relay information to higher order neural circuits (Kain and Dahanukar, 2015; Talay et al., 2017; Yapici et al., 2016).

4. Plasticity in the Fly Olfactory System

Drosophila olfactory system has emerged as a particularly useful model for studying plasticity with a tremendous array of genetic and functional tools available to dissect connectivity and assess underlying molecular pathways. Olfactory behavior is highly plastic and can be modulated by external factors like the presence of an odor and internal factors like physiological state. Modulation of olfactory behavior can be invoked by plasticity at the level of the sensory neurons as well as the mushroom body which is the higher order processing center.

5. Modulation at the Peripheral Level

5.1. Context-Dependent Modulation

Several recent studies have shed light on the mechanism underlying context dependent modulation of olfactory behavior. One remarkable example is behavioral response of the flies to CO_2. CO_2 is the key component of the "stress odor" released by the agitated flies and is detected by ab1C ORNs (Su GS et al., 2004). Activation of ab1C leads to robust aversive behavioral responses (Su GS et al., 2007). However, CO_2 is also emitted from ripe fruits that are attractive to *Drosophila*. Surprisingly, in the presence of various other ripe fruit odors, CO_2 does not trigger aversion. One effective means for flies to ignore CO_2 would be to directly suppress the response of ab1C ORNs which detect CO_2. Studies have indeed shown that 2-3 butanedione present in ripe fruit can inhibit activation of the ab1C neurons in presence of CO2, thereby making CO2 more tolerable in the presence of fruit odors (Turner SL et al., 2009).

Context dependent neuronal inhibition can also happen through lateral inhibition by another ORN in the same sensillum. Transient activation of any ORN robustly inhibits the olfactory response of its neighbor via ephatic coupling (Su CY et al., 2012). Ephatic coupling act as a neuromodulator and allows flexibility in fly's behavioral responses to CO_2. In addition to ab1C ORNs, a second level of regulation has been proposed recently. Multiple types of PNs innervate V glomerulus, the glomerulus where CO_2 sensing ORNs project (Lin HH et al., 2013). These PNs differ in their sensitivity to CO_2 and their axonal innervation patterns in the higher brain centers. Among them two PNs type are mainly responsible for flies' behavioral response toward CO_2. The response of PNv-1 to CO_2 saturates at a low concentration (0.5%), whereas PNv-2 shows graded response to different concentration of CO_2 (0.5-2%) (Lin HH et al., 2013). Segregation of the CO_2 processing circuit into multiple pathways may allow differential modulation of each individual pathway. Indeed, in lateral horn, PNv-1 input is inhibited by a specific PN types (PNv-3). PNv-3 is GABAergic and can be a parallel inhibitor pathway in the olfactory circuit (Liang L et al., 2013). It has been shown that fruit odors activate PNv-3 to selectively inhibit the PNv-1 output neurons without affecting the PNv-2 pathway that mediates aversion to high concentrations of CO_2. As a result, fruit odors may selectively inhibit the behavioral aversion to low levels of CO_2 present in ripe fruits, while the fly retains its ability to respond to high CO_2 levels that may signal danger. Much remains to be learned about how the neuronal circuit works in concert to modulate CO_2 olfactory circuit and to determine the contribution of each individual mechanism in shaping the behavioral response of flies towards CO_2.

6. Modulation of Olfactory Responses by Starvation State

Recent studies show that during starvation, cascades of neuromodulators fine tune early olfactory circuits to enhance fly's ability to detect food odors. In an elegant study, Root, Ko and colleagues show that starved flies exhibit elevated foraging efficiency. These behavioral changes require a shift in the antennal lobe odor map that includes facilitation of synaptic output from *Or42b* expressing ORNs and suppression of those from *Or85a* expressing ORNs - neuronal populations that mediate odor guided attraction and aversion behavior respectively (Root CM et al., 2011, Ko KI et al., 2015). The facilitation in *Or42b* expressing ORNs is mediated by short neuropeptide F (sNPF) (Root CM et al., 2011), a neuropeptide related to the mammalian orexigenic peptide, neuropeptide Y (NPY). Suppression of Or85a olfactory output, by contrast, is mediated by tachykinin (Ko KI et al., 2015), a fly homolog of substance P, a multifunctional neuromodulator in mammals. It has been further noticed that the expression level of sNPF receptor (*sNPFR*) and tachykinin receptor (*DTKR*) are both regulated by insulin. In the fed state, high insulin level leads to down-regulation of *sNPFR* and *DTKR* expression in *Or42b* expressing ORNs and *Or85a* expressing ORNs respectively. In contrast, during starvation, decreased insulin levels allow upregulation of *sNPFR* and *DTKR* in their respective ORNs and thus elevates the hedonic value of food odors (Root CM et al., 2011, Ko KI et al., 2015). In addition to sNPF and tachykinin, starvation likely recruits other neuropeptides such as CCHamide-1 that elevates spike responses to several ORNs (Farhan A et al., 2013).

7. Sexually Dimorphic Modulation of Olfactory Responses

Neuromodulators are also known to regulate sex drive in male flies. In a recent study it has been shown that *Or47b* expressing ORNs promote courtship in male flies by detecting an aphrodisiac pheromone, palmitoleic acid (PA) (Lin HH et al., 2016). The co-ordination of the sensitivity of the *Or47b* expressing with male fertility occurs through the action of juvenile hormone (JH). Through its action on methoprene tolerant, a putative JH receptor, JH increases the pheromone sensitivity in *Or47b* expressing ORNs in male flies and thus elevates male courtship vigor (Lin HH et al., 2016).

7.1. Plasticity in ORNs

Studies from different research laboratories demonstrate experience-dependent changes in the olfactory circuit and resultant persistent alteration in the fly's behavior. This experienced dependent modulation in the olfactory circuit is, however, restricted to a short critical period of less than a week following eclosion. A recent study by_Devaud et al., demonstrated that odor exposure during the juvenile period changes AL glomeruli architecture. Pre-exposure to specific odorants leads to reduction in odor avoidance during testing. The authors found that this plasticity in the AL and behavior is stimulus specific and restricted to a shorter period immediately following eclosion (Devaud et al., 2003). This study compared early odor exposure (2-5 days after post eclosion (dpe)) with late exposure (8-11dpe) in mature adults. Juvenile flies, when exposed to an odor, showed stronger changes whereas mature flies (i.e., late exposure) did not show any significant changes after odor exposure.

Sachse et al., (2007) described similar activity dependent plasticity in the AL by using Gal4 driver lines that specifically labelled either ORNs or PN dendrites in CO_2 responsive V glomerulus. High level of CO2 increased V glomerulus volume while not affecting surrounding glomeruli (Sachse et al., 2007, Das S 2011). CO_2 exposure must persist for 2 days to cause structural plasticity in the AL and it is reversible if high CO_2 is removed before the end of first week post-eclosion. The results obtained from Sachse et al., conflicts with earlier observation showing volume changes in the AL glomeruli is irreversible. Taken together, these studies provided evidences ruling out peripheral adaptation as a mechanism underlying olfactory experience dependent plasticity.

In disagreement with the studies described above, some studies have suggested a more important role for peripheral ORNs changes due to odor exposure during critical period. In an elegant study, Siddiqi and colleagues showed that peripheral changes in ORNs can account for olfactory behavioral plasticity. They demonstrated that exposure to a particular attractive odor during the critical period increases neuronal sensitivity to that particular odor and this increased activity of the ORNs is also reflected in an increased behavioral preference for the odor at sub-optimal concentrations (Chakraborty TS et al., 2009, Iyengar A et al., 2010). The cellular and molecular mechanisms underlying this phenotype are not yet fully understood. A recent study has shown that Notch plays a crucial role in experience-dependent plasticity (Kidd et al., 2015). Early odor experience activates non-canonical Notch signaling in ORNs, leading to increased glomerular volume, but interestingly no corresponding changes are observed in PNs. The mechanism of Notch signaling is cell autonomous in ORNs, because silencing output with tetanus toxin does not affect the glomerular volume changes in the antennal lobe (Kidd and Lieber 2016).

8. Plasticity in the Mushroom Body

Mushroom bodies, paired protocerebral neuropils in the fly brain, are known to play an important role in olfactory associative learning (Heisenberg 2003, Keene and Waddell 2007; McGuire et al., 2005). The neuronal architecture of the mushroom body has been comprehensively studied (Aso et al., 2014; Tanaka et al., 2008). The intrinsic neurons of the mushroom body are called the Kenyon cells (KCs). Dendritic arborization of Kenyon cells form a cup shaped structure called the calyx. KC axons run in parallel, forming the peduncle and finally either bifurcate into vertical (α, α') and horizontal lobes (β, β') or shape a third type of lobe, the gamma lobe (γ).

It has been shown that the neuropil of MB undergoes continuous reorganization through-out the life after being subjected to specific conditionings. A recent study showed that specific environmental input such as heat during pre-adult stage can alter MB calyx volume, but has little or no effect on other brain structure such as antennal lobe and central complex. These results show that some brain structures are especially sensitive to certain environmental factors. It is remarkable that in a fruit fly, olfactory enrichment and social contact of adults increase the number of Kenyon cell fibers in the MB peduncle (Technau, 1984). Female flies reared in high density larval cultures were shown to have more MB Kenyon cell fibers than females reared in low density larval cultures (Heisenberg et al., 1995). All these results suggest that environmentally-induced neuronal and behavioral plasticity is not limited to vertebrates, as both social and physicochemical cues affect fly brain development and function. Using in vivo Ca^{2+} imaging, recent studies show that learning related changes can be recorded in both pre and post synaptic compartments in MB. Functional studies suggest that MBONs are strong candidate for the plasticity during learning, suggesting that the MBON responses are individualized through experience dependent plasticity (Hige et al., 2015).

Neuropeptides secreted by subsets of Kenyon cells as well as those secreted from other brain neurons but acting on the KCs, can also play a crucial role in learning. sNPF has been detected in most of the KCs (Johard et al., 2008) and was shown to play a role in learning (Knapek et al., 2013). NPF is expressed in 20-26 brain neurons in fly and a subset of these appear to be presynaptic to dopaminergic neurons that innervate the mushroom body (Krashes et al., 2009). It has been shown that the NPF expression is important for formation of appetitive memory in Drosophila. Starvation enhances olfactory reward learning while well fed flies do not learn well. Stimulation of NPF neurons mimics food deprivation and promotes appetitive memory performance in fed flies (Krashes et al., 2009). This memory requires expression of the NPF receptor on a set of six dopaminergic neurons that innervate MB. These results suggest that the NPF neurons and the NPF receptor expressing dopaminergic neurons serve as a motivational switch in the MB circuits and control appetitive memory output. In addition, recently it has been shown that dopamine receptors in KCs can also play a crucial role in associative learning (Kim 2007; Qin 2012). Recent studies have revealed that DopEcR is preferentially expressed in the mushroom bodies and involved in learning. The *Drosophila* insulin receptor substrate CHICO is expressed in the Kenyon cells even in adult flies (Naganos et al., 2012) suggesting that these cells are targeted by insulin signaling. Naganos et al., 2012 also demonstrated that *Chico* mutants display defects in olfactory learning and that memory formation could be restored after *Chico* is transgenically rescued specifically in mushroom bodies. Conditional knock-down of *Chico* or *dInR* in adult Kenyon cells is required to determine acute effects of insulin signaling to the mushroom bodies in learning.

9. Plasticity in the Fly Gustatory System

The function of the gustatory system in flies goes beyond simple detection of chemical compounds. Taste preferences as well as activity of taste neurons are modulated by internal physiological states and external factors like contamination of food. This plasticity in the taste system is extremely important for flies to survive and reproduce.

Internal physiological conditions like mating for female flies or nutritional imbalance changes the fly's needs to ingest particular kinds of nutrients in higher quantities. The fly's taste preferences are likewise modulated to enable them to feed on the required diet. In natural conditions, most available food sources are mixtures of both appetitive and aversive compounds. Since it is essential for the flies to avoid toxic substances, in presence of them, the neurons signaling for appetitive responses are often modulated to prevent feeding.

10. Modulation of Taste System by External Factors

Since most aversive bitter tasting compounds are detrimental to the fly, their chances of survival as well as reproductive success depends on their ability to reject food sources contaminated with bitter compounds. Modulations of these type can occur both at the peripheral level as well as at the level of higher order neurons. One study shows that sweet neurons are inhibited when sugars are contaminated by bitter compounds (French et al., 2015). They demonstrate that it was not the outcome of ephaptic interaction, rather that of the bitter compounds directly inhibiting the sweet neurons. Whether it was through the sweet taste receptors or by altering any other property of the sweet neurons have not yet been understood. Another group has reported that an olfactory binding protein obp49a plays a significant role in inhibition of sugar neurons by bitter compounds (Zhang et al., 2013). Obp49a is secreted by the support cells into the taste sensillary hemolymph. It apparently binds with the bitter compounds and carry them to the sweet neurons. However, whether the sweet neurons are inhibited by the bitter compound or by obp49a itself is not clear. Besides bitter compounds, acids can also inhibit sweet neuron activity in flies (Charlu et al., 2013). One study using calcium imaging shows that in bitter silenced flies, inhibition of tarsal sweet neurons by bitter compounds can be inhibited by acids. There is also evidence to suggest that GABAergic interneurons play an important role in inhibition of sweet neuron activity in the presence of bitter compounds (Chu et al., 2014; Pool et al., 2014). These interneurons synapse with both sugar and bitter neurons. Bitter neuron activation leads to increased activation of the interneurons leading to increased secretion of GABA which decreases presynaptic activity of the sweet neurons via GABA-B receptors. Another piece of evidence suggests that mechanosensory neurons are also capable of inhibiting sweet neuron activity (Jeong et al., 2016). Flies generally prefer softer food over harder food even with a higher concentration of sugar. Mechanosensory neurons which sense the hardness of the food can inhibit the sweet neuron activity via GABAergic interneurons. Expectedly, this process is also dependent on GABA-B receptor.

11. Modulation of Taste by Past Experiences

A study by Zhang et al., 2013, demonstrated that components of fly taste can be modulated by prior experiences. They reported that when food is contaminated with camphor, a non-toxic bitter compound, for a long time, taste neurons are gradually desensitized to camphor. This is caused by a decline in E3 ubiquitin-ligase which down regulates Trpl, the camphor receptor in taste neurons. However, camphor sensitivity is returned to its normal level once flies are reintroduced to normal diet.

12. Modulation by Internal Physiological States

12.1. By Nutritional Imbalance

Nutritional imbalance is caused by both starvation as well as deprivation of specific macronutrients, both of which are capable of altering preferences for different foods in flies. When flies are wet-starved they have higher sensitivity towards sugar as well as lower sensitivity for bitter compounds (Inagaki et al., 2014). The enhanced sensitivity towards sugar is an outcome of increased secretion of dopamine (Inagaki et al., 2012; Marella et al., 2012). Flies fed on L-dopa, a dopamine precursor, have been shown to have increased sensitivity for sugar. Activation of the dopamine neurons by expressing the heat receptor TrpA1 also has a similar effect on sugar sensitivity. It has been reported that dopamine can act directly on the sugar neurons via DopEcR, a class of dopamine receptors expressed in the sugar neurons, to elevate sugar evoked calcium influx, increasing presynaptic activity in response to sugar stimulus (Inagaki et al., 2012). Another study reported that Dop2R function is required for increased sugar sensitivity following starvation (Inagaki et al., 2014). Interestingly, increased dopamine secretion is controlled by upstream dNPF + neurons however whether dNPF itself is required for this process has not been studied.

A different group has reported that starvation leads to up-regulation of Gr64a which could also be instrumental in increasing sensitivity to sugar (Nishimura et al., 2012). However, the role of differential regulation of GRs in the changing sugar sensitivity has not been explored. One reason might be that since their expression levels are very low, subtle differences in expression might be difficult to be understood.

The decrease in bitter neuron response upon starvation is independent of the increased sugar sensitivity (Inagaki et al., 2014). Inagaki et al., reported that inhibition of bitter neurons is mediated by sNPF, a different neuropeptide. Starvation likely triggers the lateral neurosecretory cells (LNC) to secrete sNPF which acts on sNPFRs present on GABAergic interneurons. However how GABAergic interneurons inhibit bitter compound evoked activity of the bitter neurons is yet to be understood. Lateral neurosecretory cells secrete sNPF on the Insulin producing cells (IPCs) in the brain as well. However, no role of the IPCs in decreasing bitter sensitivity could be found. Interestingly, secretion of sNPF by the LNCs is under the control of upstream *AKH*+ cells. Artificially increasing the activity of the *AKH*+ cells using *TrpA1* decreases bitter neuron sensitivity. Additionally, another study has suggested the role of the ventro-lateral cluster of octopaminergic/tyraminergic neurons (OA-VLs) in modulating bitter neurons upon starvation (LeDue et al., 2016). Starvation leads to a decrease in the neural activity in OA-VL neurons which in turn leads to secretion of octopamine and tyramine. Both octopamine and tyramine can inhibit synaptic output of the bitter neurons via Oct-TyR receptors present on the bitter neurons. Interestingly, the increase in sugar sensitivity precedes the decrease in bitter sensitivity upon increasing starvation (Inagaki et al., 2014).

Studies investigating the effect of macronutrient deprivation mostly focused on the outcome of yeast deprivation which is a natural source of proteins and amino acids in flies (Liu et al., 2015; Ribeiro and Dickson, 2010; Vargas et al., 2010). Expectedly, flies deprived of yeast have an increased preference for yeast. One study suggests that it also leads to decreased preference for sugar (Liu et al., 2017). Earlier studies postulated that TOR pathway and serotonin signaling plays significant roles in increasing yeast preference upon yeast deprivation (Ribeiro and Dickson, 2010; Vargas et al., 2010). Recently Liu et al., suggested that increased yeast intake and decreased sugar intake following yeast deprivation is controlled by a group of dopaminergic neurons. Increased yeast preference is mediated by downstream neurons expressing *Dop2R* whereas *Dop1R* expressing downstream neurons mediate decreased sugar preference. A recent study (Wang et al., 2016) found that when flies are fed with sucralose, a non-nutritive sweet compound, they have higher preference, neuronal sensitivity and neuronal responses to sugar. *Gr64a+* sweet neurons were found to be modulated in these flies. Furthermore, *InR* was upregulated in these flies. This leads to increase in insulin signaling to octopaminergic and dopaminergic neurons which ultimately cause an increase in dNPF secretion. dNPF directly acts on the NPF-receptors on the *Gr64a+* sweet neurons thereby causing an increase in sugar sensitivity. However, another group refuted this claim stating that the sucralose fed flies were essentially starved and the phenotypes observed are indistinguishable from that of wet starved flies (Park et al., 2017).

12.2. By Mating Status

In flies, egg development necessitates a larger supply of proteins and amino acids. Subsequently, preferences for yeast (natural source of proteins and amino acids for flies) and amino acids increases dramatically in female flies following mating (Ganguly et al., 2017; Ribeiro and Dickson, 2010; Vargas et al., 2010). Although sodium does not contribute towards reproductive success, interestingly, female flies also develop a higher preference for salt upon mating (Walker et al., 2015). Sex peptide, a peptide deposited by the male on the female's reproductive tract during mating plays a significant role in the post mating dietary switches (Aigaki et al., 1991; Kubli, 2003; Liu and Kubli, 2003). Some early studies on post-mating dietary switch suggested that sex peptide works directly on sex peptide receptors present on sensory neurons and detects nutrients and pheromones thereby modulating responses at a peripheral level (Kubli, 2003). Many recent studies however indicate the existence of a more sophisticated pathway. The sex peptides are detected by *Ppk+, fru+, dsx+* sex peptide sensory neurons (SPSNs) present in the reproductive tract of the female flies (Häsemeyer et al., 2009; Ribeiro and Dickson, 2010; Yang et al., 2009). The activity of the SPSNs are inhibited as an outcome. This leads to silencing of the sex peptide abdominal ganglion (SAG) neurons which connect the SPSNs to the central brain where egg-laying and mating related decisions are made (Feng et al., 2014). Octopamine secreted by a small group of sexually dimorphic *dsx+* octopaminergic neurons on the abdominal ganglion have also been postulated to mediate increased yeast preference upon mating (Rezával et al., 2014). Interestingly, modulation of salt preference following mating has been shown to be independent of octopamine (Walker et al., 2015). A role of serotonin has also been suggested since increased serotonin level leads to an increased preference for yeast (Vargas et al., 2010). But no evidence suggesting increase in serotonin level following mating has been found (Vargas et al., 2010). TOR pathway has also been found to be important in the post mating dietary switch for yeast. Activation of dS6K a component of the fly TOR pathway increases yeast preference while silencing dS6K leads to decreased appetite for yeast (Ribeiro and Dickson, 2010; Vargas et al., 2010).

Conclusion

Much has been learned about how the behavior of the fruit fly is modulated by a wide range of external as well as internal physiological factors. While physiological states like nutritional status and mating drive causes changes in behavior, it can also be altered by past experiences and via the context of the external stimulus: the exposure to odors early in life, for instance, drives plasticity in the circuit. It will be particularly important to unravel the temporal molecular requirements within the critical period. The synaptic interplay between ORNs, PNs and LNs appears complex, and it is apparent that downstream changes in the central brain, i.e., MB and LH will also participate in overall plasticity. A small number of *Drosophila* neuropeptide and peptide hormones have been analyzed for their roles in regulation of behavior and its related physiology. However, the role of neuropeptides as co-transmitters in neuron have received less attention. It is well known from studies in crustaceans and mollusks that neuropeptides and classical neurotransmitters cooperate in neuronal circuits to increase plasticity. Possibly multiple neuropeptides in the *Drosophila* antennal lobes and central complex act both as neuromodulators and cotransmitters; and indeed several recent studies have detected neuropeptides co-localizing with neurotransmitters in these brain regions. It would be interesting to see how these neuropeptides contribute to circuit regulation following changes in internal physiological state. A few examples described in this review show that few peptides (NPF/NPY) are functionally conserved from flies to humans and thus can lead to better understanding of similar phenomenon in human. The antennal lobe with its fairly well-described functional circuitry can serve for studying peptide application or interference with peptide signaling. But, the anatomical analysis of peptidergic circuitry in the CNS is still in its infancy for *Drosophila e*specially in case of gustatory system, since the upstream circuitry is still not well understood.

References

Aigaki T, Fleischmann I, Chen PS Kubli E (1991) Ectopic expression of sex peptide alters reproductive behavior of female D. melanogaster. *Neuron* 7: 557–563.

Aso Y, Hattori D, Yu Y, Johnston RM, Iyer NA, Ngo TT, Dionne H, Abbott LF, Axel R, Tanimoto H, Rubin GM (2014) The neuronal architecture of the mushroom body provides a logic for associative learning. *Elife* 3:e04577.

Chakraborty TS, Goswami SP, Siddiqi O (2009) Sensory correlates of imaginal conditioning in Drosophila melanogaster. *J Neurogenet* 23(1-2): 210-219.

Charlu S, Wisotsky Z, Medina A, Dahanukar A (2013) Acid sensing by sweet and bitter taste neurons in Drosophila melanogaster. *Nat Commun* 4: 2042.

Chu B, Chui V, Mann K, Gordon MD (2014) Presynaptic gain control drives sweet and bitter taste integration in Drosophila. *Curr. Biol.* 24: 1978–1984.

Das S, Sadanandappa MK, Dervan A, Larkin A, Sudhakaran IP, Priya R, Heidari R, Holohan EE, Pimentel A, Gandhi A, ito K, Sanyal S, Wang

JW, Rodrigues V, Ramaswami M (2011) *Proc Natl Acad Sci US* 108: E646-54.

Devaud JM, Acebes A, Ramaswami M, Ferrus A (2003) Structural and functional changes in the olfactory pathway of adult Drosophila take place at a critical age. *J Neurobiol* 56: 13-23.

Farhan A, Gulati J, Groe-Wilde E, Vogel H, Hansson BS, Knaden M (2013) The CCHamide 1 receptor modulates sensory perception and olfactory behavior in staved Drosophila. *Sci R*ep 3: 2765.

Feng K, Palfreyman MT, Häsemeye, M, Talsma A, Dickson BJ (2014) Ascending SAG neurons control sexual receptivity of Drosophila females. *Neuron* 83:135–148.

French AS, Sellier MJ, Ali Agha M, Moutaz AA, Guigue A, Chabaud MA, Reeb PD, Mitra A, Gra, Y, Soustelle L, Marion-Poll F (2015) Dual mechanism for bitter avoidance in Drosophila. *J Neurosci* 35:3990–4004.

Ganguly A, Pang L, Duong VK, Lee A, Schoniger H, Varady E, Dahanukar A (2017) A Molecular and Cellular Context-Dependent Role for Ir76b in Detection of Amino Acid Taste. *Cell Rep* 18: 737–750.

Häsemeyer M., Yapici, N, Heberlein U, Dickson BJ (2009). Sensory neurons in the Drosophila genital tract regulate female reproductive behavior. *Neuron* 61:511–518.

Heisenberg M (2003) Mushroom body memoir: from maps to models. *Nat Rev Neurosci.* 4(4): 266-275.

Hige T, Aso Y, Rubin GM, Turner GC (2015) Plasticity-driven individualization of olfactory coding in mushroom body output neurons. *Nature* 526(7572): 258-262.

Inagaki HK, Ben-Tabou de-Leon S, Wong AM. Jagadish S, Ishimoto H, Barnea G, Kitamoto T, Axel R, Anderson DJ (2012) Visualizing neuromodulation in vivo: TANGO-mapping of dopamine signaling reveals appetite control of sugar sensing. *Cell* 148:583–595.

Inagaki HK, Panse KM, and Anderson DJ (2014) Independent, reciprocal neuromodulatory control of sweet and bitter taste sensitivity during starvation in Drosophila. *Neuron* 84:806–820.

Iyengar A, Chakraborty TS, Goswami SP, WU CF, Siddiqi O (2010) Post-eclosion odor experience modifies olfactory receptor neuron coding in Drosophila. *Proc Natl Acad Sci US* 107: 9855-9860.

Jeong YT, Oh SM, Shim J, Seo JT, Kwon JY, Moon SJ (2016). Mechanosensory neurons control sweet sensing in Drosophila. *Nat Commun* 7: 12872.

Johard HA, Enell LE, Gustafsson E, Tri lie P, Veenstra JA, Nässel DR (2008) Intrinsic neurons of Drosophila mushroom bodies express short neuropeptide F: Relations to extrinsic neurons expressing different neurotransmitters. *J Comp Neurol* 507: 1479–1496.

Kain P, Dahanukar. (2015). Secondary taste neurons that convey sweet taste and starvation in the Drosophila brain. *Neuron* 85: 819–832.

Keen AC, Waddell S (2007) Drosophila olfactory memory: single genes to complex neural circuits. *Nat Rev Neurosci* 8(5): 341-354.

Kidd S, Lieber T (2016) Mechanism of Notch pathway activation and its role in the regulation of olfactory plasticity in Drosophila melanogaster. *PLoS One* 11(3): e0151279.

Kidd S, Struhl G, Lieber T (2015) Notch is required in adult Drosophila sensory neurons for morphological and functional plasticity of the olfactory circuit. *PloS Genet* 11(5): e1005244.

Knapek S, Kahsai, L, Winther AM, Tanimoto H, Nässel DR (2013) Short neuropeptide F acts as a functional neuromodulator for olfactory memory in Kenyon cells of Drosophila mushroom bodies. *J Neurosci* 33: 5340–5345.

Ko KI, Root CM, Lindsay SA, Zaninovich OA, Shepherd AK, Wasserman SA, Kim SM, Wang JW (2015) Starvation promotes concerted modulation of appetitive olfactory behavior via parallel neuromodulatory circuits. *Elife*. 24:4.

Krashes MJ, DasGupta S, Vreede A, White B, Armstrong JD, Waddell S (2009) A neural circuit mechanism integrating motivational state with memory expression in Drosophila. *Cell* 139: 416-427.

Kubli E (2003) Sex-peptides: seminal peptides of the Drosophila male. *Cell Mol Life Sci* 60: 1689–1704.

LeDue EE, Mann K, Koch E, Chu B, Dakin R, Gordon MD (2016) Starvation-Induced Depotentiation of Bitter Taste in Drosophila. *Curr Biol* 26: 2854–2861.

Lin HH, Cao DS, Sethi S, Zeng Z, Chin JSR, Chakraborty TS, Shepherd AK, Nguyen CA, Yew JY, SU CY, Wang JW (2016) Hormonal modulation of pheromone detection enhances male courtship success. *Neuron* 90(6): 1272-1285.

Lin HH, Chu LA, Fu TF, Dickson BJ, Chiang AS (2013) Parallel neural pathways mediate CO2 avoidance response in Drosophila. *Science* 340: 13338-13341.

Liu C, Bai X, Sun J, Zhang X, Li Y (2015). Behavioral Switch of Food Preference upon Sugar Deficiency Is Regulated by GPCRs in Drosophila. *J Genet Genomics* 42: 409–412.

Liu H, Kubli E (2003) Sex-peptide is the molecular basis of the sperm effect in Drosophila melanogaster. *Proc Natl Acad Sci US* 100: 9929–9933.

Liu Q, Tabuchi M, Liu, Kodama L, Horiuchi W, Daniels J, Chiu L, Baldoni D, Wu MN (2017) Branch-specific plasticity of a bifunctional dopamine circuit encodes protein hunger. *Science* 356: 534–539.

McGuire, S.E., Deshazer, M., and Davis, R.L. (2005) irty years of olfactory learning and memory research in *Drosophila melanogaster*. *Prog Neurobiol* 76: 328–347.

Marella S, Mann K, Scott K (2012) Dopaminergic modulation of sucrose acceptance behavior in Drosophila. *Neuron* 73: 941–950.

Naganos S, Horiuchi J, Saitoe M (2012) Mutations in the Drosophila insulin receptor substrate, CHICO, impair olfactory associative learning. *Neurosci Res* 73: 49–55.

Naresh Singh, R. and Nayak, SV. (1985) Fine structure and primary sensory projections of sensilla on the maxillary palp of Drosophila melanogaster Meigen (Diptera: Drosophilidae). *Int J Insect Morphol Embryol* 14: 291–597.

Nishimura A, Ishida Y, Takahashi A, Okamoto H, Sakabe M, Itoh M., Takano-Shimizu T, Ozaki M (2012) Starvation-induced elevation of taste responsiveness and expression of a sugar taste receptor gene in Drosophila melanogaster. *J Neurogenet* 26: 206–215.

Park JH, Carvalho GB, Murphy KR, Ehrlich MR, Ja WW (2017. Sucralose Suppresses Food Intake. *Cell Metab* 25: 484–485.

Pool AH, Kvello P, Mann K, Cheung SK, Gordon MD, Wang L, Scott K (2014) Four GABAergic interneurons impose feeding restraint in Drosophila. *Neuron* 83: 164–177.

Rezával C, Nojima T, Neville MC, Lin AC, Goodwin SF (2014) Sexually dimorphic octopaminergic neurons modulate female postmating behaviors in Drosophila. *Curr Biol* 24: 725–730.

Ribeiro C, Dickson BJ (2010) Sex peptide receptor and neuronal TOR/S6K signaling modulate nutrient balancing in Drosophila. *Curr Biol.* 20: 1000–1005.

Root CM, Ko KI, Jafari A, Wang JW (2011) Presynaptic facilitation by neuropeptide signaling mediates odor-driven search. *Cell* 145: 133-44.

Sachse S, Rueckert E, Keller A, Okada R, Tanaka NK, Ito K, Vosshall LB (2007) Activity dependent plasticity in an olfactory circuit. *Neuron* 56: 838-850.

Talay M, Richman EB, Snell NJ, Hartmann GG, Fisher JD, Sorkaç A, Santoyo JF, Chou-Freed C, Nair N, Johnson M., Szymanski JR, Barnea G (2017) Transsynaptic Mapping of Second-Order Taste Neurons in Flies by trans-Tango. *Neuron* 96: 783–795.e4.

Tanaka, N., Tanimoto, H., and Ito, K. (2008) Neuronal assemblies of the *Drosophila* mushroom body. *J Comp Neurol* 508: 711–755.

Technau GM (1984) Fiber number in the mushroom bodies of adult Drosophila melanogaster depends on age, sex and experience. *J Neurogenet* 1: 113-126.

Vargas MA, Luo N, Yamaguchi A, Kapahi P (2010) A role for S6 kinase and serotonin in postmating dietary switch and balance of nutrients in D. melanogaster. *Curr Biol* 20:1006-1011.

Venkatesh S Naresh Singh R (1984) Sensilla on the third antennal segment of Drosophila melanogaster meigen (Diptera: Drosophilidae). *Int J Insect Morphol Embryol* 13: 51–114.

Walker, SJ, Corrales-Carvajal, VMM., and Ribeiro, C. (2015). Postmating Circuitry Modulates Salt Taste Processing to Increase Reproductive Output in Drosophila. *Curr. Biol*. 25, 2621–2630.

Wang QP, Lin YQ, Zhang L, Wilson YA, Oyston LJ, Cotterell J, Qi Y, Khuong TM, Bakhshi N, Planchenault Y, Browman DT, Lau MT, Cole TA, Wong AC, Simpson SJ, Cole AR, Penninger JM, Herzog H, Neely GG (2016) Sucralose Promotes Food Intake through NPY and a Neuronal Fasting Response. *Cell Metab*. 24: 75–90.

Yang CH, Rumpf S, Xiang Y, Gordon MD, Song W, Jan LY, Jan YN (2009) Control of the postmating behavioral switch in Drosophila females by internal sensory neurons. *Neuron* 61: 519–526.

Yapici N, Cohn R, Schusterreiter C, Ruta V, Vosshall LB (2016) A Taste Circuit that Regulates Ingestion by Integrating Food and Hunger Signals. *Cell* 165: 715–729.

Zhang YV, Raghuwanshi RP, Shen WL, Montell C (2013) Food experience-induced taste desensitization modulated by the Drosophila TRPL channel. *Nat Neurosci* 16: 1468–1476.

In: Synaptic Plasticity
Editor: Sraboni Chaudhury
ISBN: 978-1-53613-606-7

Chapter 4

SYNAPTIC PLASTICITY AND ELECTRICAL ADAPTATION IN NEURONS

***Rakesh Kumar**, PhD**
Division of Molecular Neurobiology,
Department of Medical Biochemistry and Biophysics,
Karolinska Institute, Stockholm, Sweden

ABSTRACT

The brain is made of millions of neurons, which upon exposure to environmental cues or with experiences, adjust their connections and circuitry, thereby leading to a desired thought, behavior and feeling. Synaptic plasticity is the mechanism by which experience rearranges neural activity, involving the strengthening or weakening of existing synapses. It programs the synaptic facilitation and depression to strengthen the synaptic transmission at pre-existing synapses. Transient forms of synaptic plasticity have been associated with temporary adaptations to sensory inputs, minor changes in behavioral states and short-lasting memory. Long lasting changes are supposed to play important roles in the construction of neural circuits during development and long-term memory. Beside structural modifications, electrical

* E-mail: rakeshpandey12@gmail.com; Mobile No: +46731951463.

adaptations also occur in neurons, including activity dependent change in the synaptic communication, either by enhanced (long-term potentiation, LTP) or depressed (long-term depression, LTD) neural functions. These changes can exist from milliseconds to long-term modifications that may persist from days or weeks and perhaps even longer. The LTP possibly increases spine density as well as the chance of multiple-spine synapses communication, where two adjacent spines arising from the same dendrite contact a single presynaptic bouton. However, the LTD involves a shrinkage of postsynaptic density (PSD) and probably initiating a complete loss of dendritic spine and its corresponding presynaptic bouton. Plethora of imaging, electrophysiological and behavioral work advocate the possible forms of synaptic plasticity and their underlying mechanism. Elucidation of electrophysiological adaptions during synaptic plasticity and its role in behavioral wellbeing at one place will have profound implications in understanding the neural basis of cognition and behavior. This chapter aims to broaden the current understanding of synaptic plasticity and how maladaptive synaptic plasticity may contribute to neuropsychiatric disorders.

Keywords: plasticity, electrophysiology, LTP, LTD, behavior, memory

1. INTRODUCTION

The brain is a structure responsible for what we call the mind. It is a seat for thinking, feeling, wanting, perceiving, curiosity, behavior, learning and memory. From the simple task to complex physical, chemical, and social cues induce rearrangement of neural circuitry via modifications in synaptic transmission. The “brain is plastic” refers to physical changes continuously taking place at the level of microscale connections between neurons. These changes in neuronal connections are the primary mechanism for learning and memory and are known as “synaptic plasticity.” Synapses transmit information via dendrites to the soma to trigger output via axonal action potentials, while dendrites help to determine synaptic plasticity, the synaptic activity in turn regulates dendritic excitability therefore, an activity-dependent reciprocal loop between synaptic plasticity and dendritic excitability exists. Although dendritic morphologies are typically relatively static, their electrical properties can change in an activity-dependent manner.

The nervous system is genetically hard-wired but neural circuits undergo extensive sculpting and rewiring in response to a variety of stimuli. Synaptic plasticity specifically refers to the activity-dependent modification of the strength or efficacy of synaptic transmission at preexisting synapses. The vast majority of synaptic inputs impinge on the dendrites rather than the soma or the axon. Therefore, dendrites serve an important role in the integration of information delivered by the synapses. Postsynaptic activity is profoundly determined by the properties of dendritic arbor itself, implying that the dendrites must in part determine the rules that govern synaptic plasticity. It follows that the type of dendritic excitability associated with a synapse should help to regulate the plastic properties of that synapse. Similarly, since the dendrite transforms the electrical and chemical signals it conveys, the location of a synapse in the dendritic tree could influence the plasticity rule (Sjöström et al., 2008). The dendritic tree of a typical neuron receives tens of thousands of inputs exhibiting different strengths and activity patterns (Braitenberg and Schüz, 1999). Prolonged activity brings about downregulation of the synaptic gain, whereas too little activity results in increased gain. Mechanistically, the spine Ca^{2+} signaling is considered fundamental in determining the sign and strength of long-term synaptic plasticity (Sjöström et al., 2008).

The synaptic plasticity signaling pathway engage changes in synaptic and cell intrinsic physiology. For example, specific circuits in the somatosensory and visual cortex involves cell type-specific structural plasticity: some buttons and dendritic spines appear and disappear, accompanied by synapse formation and elimination, respectively (Holtmaat and Svoboda, 2009). Furthermore, the olfactory bulb neuron axons undergo continuous activity-dependent synaptic remodeling that persists into adulthood (Cheetham et al., 2016). The physiological adaptations related to motor learning require neural networks to detect correlations between environmental events and to store this information as changes in synaptic strength.

Historically, it was believed that the number of neurons in the brain remained stable throughout life. The neuroanatomist Santiago Ramon y Cajal proposed that memories must therefore be formed by the strengthening of existing neuronal connections (Cajal, 1894). Later, the psychologist Donald Hebb, suggested that neurons that "fire together, wire together," which is to say that if two cells consistently fire at the same time, then the strength of the connection between them will grow stronger (Hebb, 2005). In 1970, Eric Kandel's group established a clear connection between the physical changes that occurs at individual synapse and behavioral evidence of learning and memory (Castellucci et al., 1970). In 1973, evidence for similar type of observation was reported by Bliss and Lomo, who found Long-lasting potentiation of synaptic transmission in the dentate area following stimulation of the perforant path (Bliss and Lomo, 1973). The most extensively studied and therefore prototypic forms of synaptic plasticity are the LTP and LTD observed in the CA1 region of the hippocampus, which are triggered by activation of N-methyl-D-aspartate (NMDA) receptors (NMDARs). More recently, Masao Ito and his colleagues identified a similar but distinct form of synaptic plasticity in the cerebellum known as long-term depression or LTD (Ito et al., 1982). Their investigations shows that the electrical stimulation of the cerebellum produced a long-lasting decrement in synaptic efficacy. Like hippocampal LTP, cerebellar LTD is Hebbian and exhibits properties typical of behavioral processes (Ito, 1989). Because both, hippocampus (Zola-Morgan and Squire 1990) and the cerebellum (Thompson, 1990) are important for various forms of behavioral processes in mammals, hippocampal LTP and cerebellar LTD are considered to be strong candidates for the locus of synaptic change in learning, while the associative cortical areas are thought to be the center for permanent synaptic change, after the learning has taken place (Chaillan et al., 1998; Roman et al., 1987; Squire and Zola 1996). In the 1970s and 1980s, experiments in invertebrate model systems, such as Drosophila melanogaster and *Aplysia californica*, provided compelling data to show that short-term memory is mediated by transient posttranslational modifications, particularly phosphorylation by protein kinases (Kandel and Schwartz, 1982; Dudai, 1988).

In the past few years, a candidate enzyme PKMzeta, an isoform of mammalian protein kinase C (PKC), which is expressed exclusively in neural tissue and enriched in the forebrain (Hernandez et al., 2003; Sacktor, 2008) has emerged as memory engine. This molecule is the engine of memory, constantly whirring to store information in our brains. When we learn new things, PKMzeta shows up at the gaps between neurons (synapses) and boosts the signals that go across them. Under basal conditions, dendritic PKMzeta mRNA is translationally repressed. Induction of learning-related synaptic enhancement, such as long-term potentiation (LTP) (Malenka and Bear, 2004), causes the release of the translational block on PKMzeta (Kelly et al., 2007; Muslimov et al., 2004). A synapse with PKMzeta is twice as strong as it otherwise would be, but once our memories shift to long-term storage, PKMzeta is one of a few molecules that can reinforce them, and the only one that works in healthy individuals (Shema et al., 2011).

In 2007, Shema et al., showed that blocking PKMzeta by injecting ZIP (a chemical that neutralizes PKMzeta) disrupts all sorts of memories from detection of artificial sweetener path, old fears to motor skills to the location of objects (Shema et al., 2007). The PKMzeta works by increasing the levels of α-amino-3-hydroxy-5-methyl-4-isoxazolepropionic acid receptor (AMPAR), a protein that sits at synapses and allows fast signals to travel across them. Other proteins try to drive AMPAR away from the synapse, so PKMzeta has to fight to keep it there. Thus, it is easy to erase memories with ZIP. If a synapse is devoid of PKMzeta, the tide of battle turns causing AMPAR driven synapse weakening and the memories are forgotten (Migues et al., 2010). These observations are promising but important questions about PKMzeta and its role in memory maintenance remain elusive. In addition, the underlying mechanisms could vary depending on the circuits in which they function. Even long-term memories are constantly on the verge of being erased and electrophysiological and molecular manipulations needed to keep them stable are not well defined. A key question remain, what are the molecular processes enables development of activity-dependent adaptions like LTP and LTD? This chapter aims to bring molecular and electrophysiological observations thus far on one place to widen the understanding of synaptic plasticity.

2. Memory and Synaptic Plasticity

Memory is one of the most fundamental mental processes. The neural populations interact during development to form behaviorally important neural circuits and to elucidate neurobiological correlates of perceptual and motor predispositions. Without memory, we are capable of nothing but simple reflexes and stereotyped behaviors. The plastic change often results due to alteration in the number of neurotransmitter receptors located on a synapse (Gerrow and Triller, 2010). Synaptic transmission can be either enhanced or depressed by activity, and these changes span temporal domains ranging from milliseconds to hours, days, and presumably even longer. Most forms of short-term synaptic plasticity are triggered by short bursts of activity causing transient rearrangements in presynaptic nerve terminals.

The striatum, Hippocampus and cerebellum are the most electrophysiologically explored brain areas in combination with the behavioral performance, to understand synaptic plasticity induced memory formation. Striatum controls multiple cognitive aspects including motivation, reward perception, decision making and motor planning (Hernandez et al., 2016). Striatal circuitry mediates procedural or implicit learning. Within the striatum, two forms of long-lasting synaptic plasticity have been described at the glutamatergic cortico-striatal synapses, namely long-term depression (LTD) and long-term potentiation (LTP) (Ding et al., 2008). Furthermore, basolateral amygdala (BL) has long been associated with emotion and motivation, playing an essential role in processing both positive (Tye et al., 2010) and negative (Rumpel et al., 2005) emotion-associated events, while the hippocampus is essential in the formation of spatial learning and memory (Ruediger et al., 2012). One of the best characterized forms of LTD occurs in the synapses between granule cells and Purkinje neurons in the cerebellum (Jörntell and Hansel, 2006).

3. UNDERSTANDING SYNAPTIC PLASTICITY THROUGH ELECTROPHYSIOLOGICAL OBSERVATIONS

Whole-cell recordings, and patched spines of neurons could be used to induce LTP at particular synapse, avoiding disruption of the intracellular milieu critical for LTP (Weber et al., 2016; Matsuzaki et al., 2004). An alternate approaches to studying LTP includes the use of sharp electrodes, perforated patch, spike-timing dependent plasticity, or extracellular recordings under specific conditions such as magnesium-free solution (Calabresi et al., 1992; Charpier and Deniau, 1997; Fino et al., 2005; Shen et al., 2008). It has been reported that high frequency stimulus (HFS) can induce either LTD or LTP in slices, depending on stimulating electrode placement, striatal sub-region, and age of animal (Partridge et al., 2000; Spencer and Murphy, 2000). The changes in population spike amplitude and frequency could reflect modifications in cell excitability induced by the training procedures. However, it does not necessarily represent that the synapses are altered by the conditioning procedures. For example, the population spike amplitude increase could be due to a change in the synaptic efficacy of perforant path (projections from the entorhinal cortex to the dentate gyrus) connections as a result of behavioral conditioning or it could be due to a tonically altered input to the dentate gyrus from one of its afferents. Experiments demonstrate that behavioral performance could bring an increase in the population spikes, resembling enhanced synaptic efficacy in the hippocampus and particularly in the dentate gyrus.

Rats with chronically implanted stimulating and recording electrodes in the hippocampus could be subjected to behavioral experiments consisting of reward behavior-dependent changes in the evoked potentials. The frequency and strength of the simultaneously recorded theta activity in CA1 and the dentate gyrus varies in opposite way. The perforant path stimulation during the negative-going phase of the dentate theta cycle evokes significantly greater granule cell response than stimuli during the positive-going phase (Buzsàki et al., 1981). Next, during a footshock-motivated discrimination task, the granule cells of the hippocampus develop functional changes that are similar or identical to the LTP phenomenon. In this study, the monosynaptically evoked response of the dentate area was recorded following stimulation of the medial entorhinal cortex both before and after the training sessions. The EPSP component, which is the sign of the locus for synaptic change, showed no reliable change. However, the population spike became larger. The enhanced population spike was not present immediately after the completion of training but developed in a rather gradual manner (RuÈthrich et al., 1982).

In an early study (Berger et al., 1976), demonstrated a rapid and substantial increase in hippocampal pyramidal cell activity during nictitating membrane response conditioning. The perforant path was stimulated and dentate field potentials elicited were recorded during nictitating membrane response conditioning, in order to test changes in the granule cell responsiveness (Weisz et al., 1984). Above experimental observations suggests that, the population spike amplitudes increased over pretraining baseline values during the course of training in paired conditioning animals but not in unpaired control animals, Animals transferred from one environment to another exhibited substantial increases in population excitatory postsynaptic potentials (EPSPs) and large decreases in both the amplitude and the peak latency of population spikes. Meaning that, the exploratory behavior is accompanied by a substantial increase in evoked synaptic efficacy in granule cells, an increase that long outlasts the behavior that produces it (Green et al., 1990; Croll et al., 1992). Another key experiment explored the efficacy of synaptic transmission from the perforant path to the granule cells in the dentate gyrus was monitored electrophysiologically over the course of training in an appetitively motivated, discriminated operant paradigm. After the behavioral sessions, evoked potentials were recorded from the dentate gyrus following stimulation of the perforant path over range of currents, and the amplitudes of the population spikes were measured. Significant increases in population spikes were observed after training. The increase in granule cell responsiveness persisted for at least 10 days following the cessation of behavioral trials and was in many ways comparable to LTP (Skelton et al., 1987).

Later, to overcome the methodological difficulties of recording electrophysiological parameters for synaptic plasticity in moving animals, brain slices of 300-500 um thickness were used. For the first time pseudoconditioned and conditioned rabbits were sacrificed and slices of dorsal hippocampus were made (LoTurco et al., 1988). Profile of the peak post-synaptic potential amplitudes across a train of electrical stimulations (300 msec at 5 Hz) of the Schaffer collaterals was enhanced in hippocampal CA1 pyramidal cells from rabbits that were classically conditioned relative to cells from naive and pseudoconditioned animals. The enhanced synaptic responsiveness in cells from conditioned animals and reduced after-hyperpolarization contribute together to the functioning of hippocampal CA1 pyramidal cells during classical conditioning.

In other experiments, high-frequency stimulation was applied to the perforant path during rapid-eye-movement sleep after two-way active avoidance training (Bramham et al., 1994). Plasticity decreases just after learning and restored during the stages of paradoxical sleep. This finding is consistent with the view that significant information learned during the arousal stage will be reactivated and stabilized during paradoxical sleep (Bloch and Laroche, 1984; Hars et al., 1985; Hennevin et al., 1993). Interesting behavioral and electrophysiological data using the avoidance response rates in a shuttle avoidance paradigm was reported, establishing the learning ability of rats and the facilitation of synaptic efficacy (Ramírez et al., 1988; Ramirez and Carrer, 1989). After the conditioning session, these animals were killed and the frequency threshold needed to induce LTP was determined in transverse hippocampal slices. It yielded a negative correlation between the conditioned response rate in the shuttle box and the threshold frequency required to enhance synaptic efficacy in the dentate gyrus following tetanic stimulation of the perforant path.

Also, an increase in synaptic efficacy in the dentate gyrus cells was observed during a classical conditioning task where, rats learned to make the association between a conditional stimulus (CS) (a tone) and an unconditional stimulus (US) (an electrical shock) (Doyère et al., 1993; Laroche et al., 1991). The selective increase in hippocampal cell activity during the presentation of the CS was observed (Berger and Thompson, 1977; Laroche et al., 1983). This early involvement of the hippocampal circuitry occurred in parallel with the learning of the CS - US associations (Berger et al., 1976; Laroche et al., 1987) and seems to be related to the central representation of the integrated information (Laroche et al., 1991; Laroche et al., 1987; Neuenschwander-El Massioui et al., 1991).

In order to saturate synaptic efficacy by reaching the saturation level of the LTP, a high-frequency electrical stimulation is needed. High frequency stimulus bursts on perforant path fibers shows a substantial increase in both the EPSP and the population spike in the dentate gyrus (Castro et al., 1989). At the peak of synaptic efficacy saturation, the rats got severely impaired for spatial learning, as indicated by their failure to show differential searching approach during the probe trial and longer escape latencies during training. However, following the decay of the synaptic efficacy, there were no differences across groups in either probe quadrant search time or escape latency (Castro et al., 1989). These results support the general hypothesis that spatial memory involves an initial, temporary synaptic encoding of information in the fascia dentata, some of which may be destined for more permanent storage at subsequent processing stages (Castro et al., 1989). Lately, it has been suggested that abnormal electrical activity induced by high-frequency stimulation of the perforant path may be responsible for the disruption of spatial learning, previously attributed to LTP in synaptic efficacy saturation (McNamara et al., 1993).

3.1. Electrical Stimulation and Behavioral Consequences of Synaptic Plasticity

When high-frequency electrical stimulation (each trains of 400 Hz for 20 msec separated by 15 sec) was given on the perforant path (in naive implanted rabbits) for 5 days in succession, all animals exhibited significantly enhanced population spike amplitudes after the first series of high frequency trains (Berger, 1984). The behavioral results showed that animals in which synaptic changes were produced by the stimulation, learned the discrimination task significantly faster than control animals. Thus, the data support the assumption that long-lasting enhancement of synaptic efficacy increases the rate of behavioral learning in a simple discrimination task. Furthermore, high-intensity stimulation of the mesencephalic reticular formation (m.r.f.) had a facilitating effects on the development of associative changes in hippocampal and dentate during classical conditioning (Laroche et al., 1983; Bloch, 1970; Bloch and Laroche 1981). On the other hand, the slow-intensity m.r.f. stimulation facilitated long-term enhancement of synaptic efficacy compatible with behavioral findings. There are extensive literatures suggesting an association between negative emotion and cognitive impairment. The basolateral amygdala (BL) has long been associated with emotion and motivation, playing an essential role in processing both positive (Tye et al., 2010) and negative (Rumpel et al., 2005) emotion-associated events, while the hippocampus is essential in the formation of spatial learning and memory (Ruediger et al., 2012). Recently, helplessness (LHL) and learned hopefulness (LHF) by exposing animals to inescapable foot shocks or with anticipated avoidance trainings. The LHF animals showed spatial memory potentiation with excitatory monosynaptic upscaling between posterior basolateral amygdale (BLP) and ventral hippocampal CA1 (vCA1), whereas the LHL showed memory deficits with an attenuated BLP–vCA1 connection. The excitatory connection between BLP and vCA1, and opposite scaling of BLP–vCA1 inputs control hopefulness- and helplessness-modulated spatial learning and memory (Yang et al., 2016).

3.2. Frequency Dependent Synaptic Potentiation and the Role of Glutamate and GABA Receptors

Appropriate glutamate dynamics for the optimal temporal contingency between pre- and postsynaptic activity is required for spike timing-dependent plasticity (STDP) emergence. Aberrant non-timing-dependent plasticity occurs for the uncorrelated events (Valtcheva and Venance, 2016). However, excessive expression of these forms of plasticity can be detrimental for behavior and motor activity. The loss of synaptic downscaling alters LTP regulation in the basal ganglia, as well as in the cortex, leading to hyperkinetic movement disorders (Calabresi et al., 2016). Synapses with a low initial probability of release function as high-pass filters, since they get facilitated during high-frequency action potential bursts while low-frequency bursts are not transmitted with the same efficacy. In contrast, synapses with a high initial probability of release, function as low-pass filters and they are depressed during high-frequency bursts to relay low-frequency activity (Abbott and Regehr, 2004). Activity induced long-term modification of glutamatergic synapses depends on the frequency of synaptic activation. The N-methyl-D-aspartate receptor (NMDAR) mediated Ca^{2+} signals in individual spines are highly sensitive to the spatiotemporal activity pattern of even a few nearby synapses (Weber et al., 2016). The cooperative amplification of spine Ca^{2+} signals are produced by a graded NMDAR-mediated mechanism (Major et al., 2013), most likely due to effective propagation of EPSPs between adjacent spines in distal compartments, alleviating the Mg^{2+} block of NMDARs and leading to direct Ca^{2+} influx in coactive nearby spines. In high-impedance terminal, small spine-to-dendrite voltage attenuation produces strong dendritic depolarization (Harnett et al., 2012; Rinzel and Rall, 1974), allowing even a few closely located inputs to interact, whereas larger numbers of inputs are necessary to evoke nonlinear amplification of synaptic Ca^{2+} signals at low-impedance proximal dendritic locations. Indeed, the same mechanism affects dendritic spike properties depending on the input location and spatial pattern. In most experiments, EPSPs were measured at an additional nearby (<15 μm) reference spine that are not stimulated during the cooperative LTP protocol.

The long-term modification of developing rat hippocampal GABAergic synapses induced by repetitive coincident pre and postsynaptic spiking is also frequency dependent. Spiking at 20-50 Hz results in synaptic potentiation, whereas spiking at 5 Hz leads to synaptic depression

(Xu et al., 2008). The potentiation is abolished by blocking GABAB receptors (GABABRs), whereas the depression is independent of GABABR activation and could be converted to potentiation by elevating GABABR activity. The activity of postsynaptic Ca^{2+}/calmodulin dependent protein kinase II is necessary for LTP in developing GABAergic synapses and its phosphorylation at Thr286 could be enhanced by activating GABABRs with baclofen (Giese et al., 1998). Together with the finding of frequency dependent activation of GABABRs, these results indicate that postsynaptic GABABR activation mediates frequency dependent potentiation of developing GABAergic synapses. Also, the potentiation could be attributed to a local postsynaptic increase in Na(+)/K(+)/2Cl(-) co-transporter activity near activated synapses(Xu et al., 2008). GABABR signaling can enhance alpha-CaMKII phosphorylation at Thr286.

4. Molecular Mechanism of Long-Term Plasticity Induction

Long-term potentiation (LTP), by definition is long-lasting, and this longevity of LTP is possibly essential for its role as an information storage mechanism. Bliss and Lomo 1973, demonstrated that the excitatory postsynaptic potentials (EPSPs) evoked by electrical stimulation in the dentate gyrus of the rabbit hippocampus increased in amplitude after repeated high-frequency stimulation, a phenomenon termed as LTP. Long-term potentiation LTP of synaptic activity is by far the most popular and widely researched model of synaptic plastic changes that might occur during learning. LTP underlies learning and memory in some mammalian brain regions, in particular the amygdala (Maren, 2001; Maren, 2005; Rodrigues et al., 2004). The important role of LTP in learning and memory in mammals is continuously growing, even in complex brain regions such as the hippocampus (Moseret al., 1998; Whitlock et al., 2006) and neocortex (Rioult-Pedotti et al., 1998; Rioult-Pedotti et al., 2000). LTP is frequency-dependent (Kirkwood et al., 1993; Sjöström et al., 2001) and requires *N*-methyl-D-aspartate (NMDA) receptor-mediated Ca^{2+} influx (Schiller et al., 1998; Neveu and Zucker, 1996; Malenka et al., 1988; Magee and Johnston, 1997). Furthermore, the correlated activity leads to synaptic strengthening such that the presynaptic cell drives the postsynaptic cell more reliably. LTP also exhibits at least two phases, denoted as early and late LTP (Kandel, 2001; Reymann and Frey 2007). The early phase of LTP lasts only 1–3 h and does not require protein synthesis, whereas the late phase lasts for at least a day and requires both translation and transcription (Kandel, 2001; Reymann and Frey 2007). Thus, a key concept is that synaptic strength at excitatory synapses is bi-directionally modifiable by different patterns of activity. External cues translate the calcium signal required to trigger LTP into the long-lasting increase in synaptic strength (Sanes and Lichtman, 1999; Malenka and Bear, 2004). The calcium/calmodulin (CaM)-dependent protein kinase II (CaMKII) fulfills these requirements and act as a key component of the molecular machinery for LTP. CaMKII undergoes autophosphorylation after the LTP trigger (Barria et al., 1997; Fukunaga et al., 1995), and LTP induction gets prevented both in knockout mice lacking a critical CaMKII subunit (Silva et al., 1992), and in knock-in animals in which endogenous CaMKII is replaced with a form lacking the autophosphorylation site Thr286 (Giese et al., 1998). The extracellular signal-regulated kinase (Erk)/mitogen-activated protein kinase (MAPK) pathway has also been suggested to be important for LTP, as well as some forms of learning and memory (Sweatt, 2004; Thomas and Huganir, 2004). In addition, Src kinase has been implicated in the enhancement of NMDAR function during LTP induction (Kalia et al., 2004). Moreover, Cdk5 inhibition (Pharmacological, brain-wide Cdk5 conditional knockout or viral-mediated specific loss of Cdk5) impairs dopamine-facilitated LTP in striatum (Hernandez et al., 2016). Treatment with a GSK3β inhibitor during the juvenile stage, effectively rescue dendritic spine and working memory deficits in adult animals. Glycogen synthase kinase-3β (GSK3β) activity affects the LTP/LTD balance in prefrontal cortical neurons, thus providing a mechanistic interpretation of how GSK3β hyperactivity affects synaptic instability and resultant spine loss in an animal model for Schizophrenia

(Xing et al., 2016). As discussed earlier, the atypical PKC isozyme, PKMz has received much attention in the last decade because this isozyme is rapidly expressed upon induction of LTP and called as memory engine.

5. Long-Term Depression (LTD)

In term of neurophysiology, LTD is an activity-dependent reduction in the efficacy of neuronal synapses lasting hours or longer. The weakening of a synapse is necessary to optimize information storage in a neural network. It can be both homosynaptic and heterosynaptic. Electrophysiological evidence for heterosynaptic LTD was found in the hippocampus a few years after the publication of Stent's conjecture (Stent, 1973); attempting to induce LTP by high-frequency stimulation of one pathway produced LTD in other, inactive pathways. Prolonged low-frequency stimulation (LFS; typically, 900 pulses at 1–4 Hz) results in LTD. In other words, homosynaptic and heterosynaptic LTD are both the functional inverse of the Hebbian postulate, but in different manners.

One of the best characterized forms of LTD occurs in the synapses between granule cells and Purkinje neurons in the cerebellum (Jörntell and Hansel, 2006). The induction of LTD in cerebellum involves simultaneous stimulations of parallel fibres and climbing fibres at low frequency (~1 Hz), which promote large elevations of the intracellular calcium ion concentration (Coesmans et al., 2004), to activate a positive feedback loop. Mechanistically, the activation of this loop is regulated by protein kinase C (PKC), cytosolic phospholipase A2 (PLA2), and the extracellular signal-regulated protein kinase (ERK) pathway activation (Tanaka and Augustine, 2008) causing reduction in the population of synaptic AMPA receptors. LTD induction results into the phosphorylation of synaptic AMPA receptors (AMPARsyn) by PKC (Chung et al., 2003), which disrupts their interactions with the glutamate-receptor interacting protein (GRIP) (Henley and Wilkinson, 2013), to causes their endocytosis (Anggono et al., 2013). Importantly, it is well established that most synapses that exhibit LTP also express one or more forms of LTD. By far, the most commonly reported form of cortico-striatal plasticity is LTD that can be induced in response to high-frequency stimulation (HFS) *in vitro* (Gerdeman et al., 2002). The induction of striatal LTD requires postsynaptic depolarization and endocannabinoid release (Calabresi et al., 1992).

6. Synaptic Enhancement and Downscaling

The balance between excitatory and inhibitory synaptic transmission is sought for the adequate and proper processing of inputs by brain and alterations in this balance can cause circuit dysfunction and disease. The processing of information relies on change in the strength of connection between neurons. Short-term synaptic enhancement results from an increase in the probability of synaptic terminals releasing transmitters in response to pre-synaptic action potentials. Synapses get strengthen for a short time because of either an increase in the size of readily releasable pool of packaged transmitter or an increase in the amount of packaged transmitter released in response to each action potential (Stevens and Wesseling, 1999)· Depending on the time scales over which it acts synaptic augmentation is classified as synaptic facilitation or post-tetanic potentiation.

Synaptic downscaling represents the decrease in strength of a synapse to reverse synaptic plasticity and an associated increase in the excitability of excitatory neuronal. The loss of synaptic downscaling alters LTP regulation in the basal ganglia, as well as in the cortex, leading to hyperkinetic movement disorders. The downscaling of plasticity in striatal glutamatergic synapses on striatal neurons if impaired leads to early cognitive and motor symptoms (Calabresi et al., 2016). The loss of downscaling in both cortical and subcortical structures might represent a common synaptic mechanism occurring in distinct hyperkinetic disorders, as detected using neurophysiological approaches. Individual synaptic strengths could be scaled up or down globally, to counteract decreased or increased levels in synaptic drive. Evidence for such synaptic scaling in the mammalian central nervous system was first discovered by Turrigiano et al., 1998 and Desai et al., 2002 in neocortical cell cultures. In synaptic scaling, either the presynaptic release machinery can be regulated (Paradis et al., 2001; Burrone et al., 2002) or the postsynaptic sensitivity to neurotransmitter can be altered (Turrigiano et al.,1998; Goel et al., 2006; Okabe et al., 1999). Since striatum is the primary site of neurodegeneration, it has been hypothesized that downscaling of the plasticity of glutamatergic synapses of Striatal projection neurons (SPNs) to reverse synaptic strength to pre-LTP levels disrupts regulatory processes that destabilize neuronal circuits during information storage and contributes to the impaired behavioral flexibility described in early-stage huntington disease (HD) (Yang et al., 2016).

7. Synaptic Homeostasis

It is phenomenon of recalibration or scaling the synaptic strength to increase the net synaptic strength and cortical excitability along with decreased inducibility of associative synaptic long-term potentiation (LTP). The major form of homeostatic plasticity is 'synaptic scaling,' where the strength of all synapses on a given cell is adjusted in response to prolonged changes in activity (Turrigiano and Nelson, 2004). Specifically, prolonged decreases in overall activity cause a net scaling up of total synaptic strengths while prolonged increases in activity cause a scaling down of synaptic strengths. This supports the idea of having reduced LTP-like plasticity in the morning and facilitated LTP-like plasticity in the evening in healthy humans (Sale et al., 2008). Some evidence indicates that the induction of LTD is facilitated after sleep deprivation in the rat hippocampus (Tadavarty et al., 2009). The inducibility of LTP is attenuated both immediately after sleep (insufficient upscaling) and after extended wakefulness (excessive upscaling and saturation). Markers of synaptic strength show a wake-dependent rise and a sleep-dependent decline, such as the slope of the local field potential elicited by electrical stimulation in the rat cortex (Vyazovskiy et al., 2008). Sleep–wake dependent homeostatic modifications appear to refine the set-point for the induction of Hebbian long-term potentiation (LTP) of glutamatergic synaptic transmission, a key mechanism of associative plasticity and a molecular correlate for learning and memory (Bliss and Collingridge, 1993). Sleep recalibrates homeostatic and associative synaptic plasticity because the induction of LTP using high frequency electrical stimulation in the motor cortex of rats is partially occluded after prolonged wakefulness and restored after sleep (Kuhn et al., 2016). Furthermore, a significant deficit in the encoding of declarative memory occurs after sleep deprivation (word-pair task), which is thought to depend on the induction of LTP in a hippocampal-neocortical network (Kuhn et al., 2016; Eichenbaum, 2004). This form of plasticity or scaling operates on a much slower timescale than LTP or LTD and may be particularly important during the development of neural circuits. At the level of synapse, excitatory inputs are typically downregulated by extended high-activity periods, whereas upregulation is brought about by episodes of low activity to bring synaptic homeostasis. Whether primarily the pre- or the postsynaptic side is homeostatically regulated may depend on developmental stage (Wierenga et al., 2005; Wierenga et al., 2006). In some cases, both the pre- and the postsynaptic sides are modified simultaneously (Wierenga et al., 2006; Thiagarajan et al., 2005). Also, the homeostasis of intrinsic excitability factors in neurons may occur raising the possibility that individual sections of the dendritic tree might be regulated separately.

9. Short-Term Plasticity

The ratio of excitatory and inhibitory (E/I ratio) synaptic transmission is not static, but can be modulated by synaptic plasticity. Temporal pattern of activation regulates the strength of both excitatory and inhibitory synapses to cause short-term plasticity (Davies et al., 1990; Zucker and Regehr, 2002; Bartley and Dobrunz, 2015). A single neuron can contact multiple postsynaptic partners using synapses with different forms of short-term plasticity, suggesting that the short-term plasticity is tuned in a target-dependent manner (Rozov et al., 2001; Bao et al., 2010). Synapses preferentially transmit certain patterns of presynaptic activity as a consequence of their short-term plasticity. Synapses with a low initial probability of release, function as high-pass filters, since they get facilitated during high-frequency action potential bursts while low-frequency bursts is not transmitted with the same efficacy. In contrast, synapses with a high initial probability of release function as low-pass filters, since they get depressed during the high-frequency bursts to relay low-frequency activity (Abbott and Regehr, 2004). In this way, facilitation of synapses may act as high-pass filters (Atluri and Regehr, 1996) and could play a special role in encoding burst activity that is a common feature of neural circuits (Lisman, 1997). In contrast to facilitating synapses, depressing synapses are low-pass filters that preferentially signal the onset of activity after periods of quiescence (Abbott et al, 1997; Rose and Fortune, 1999). Synapses where facilitation and depression are both present can act as band-pass filters (Dittman et al., 2000). Short-term plasticity could also be important for the induction of long-term plasticity or facilitation of neuronal synapses. Besides, the strength of inputs onto inhibitory interneurons is also modulated (Wierenga and Wadman, 2003; Sun et al., 2005; Bartley et al., 2008), potentially altering inhibition through changes in interneuron recruitment (Bartos and Elgueta, 2012; Kispersky et al., 2012). The regulation of the E/I ratio by short-term plasticity is therefore a complex process. Furthermore, involvement of Glia in the Short-Term Plasticity has been documented; glia may sense the extracellular messengers and then release substances that in turn can directly affect synaptic efficacy (Haydon, 2001; Araque et al., 2014). For example, glia expresses many different neurotransmitter receptors (eg, glutamate receptors), which when activated result in the release of substances (eg, ATP) that can then act on presynaptic terminals to regulate neurotransmitter release.

10. Synaptic Facilitation

Synaptic facilitation is a phenomenon in which postsynaptic potentials (PSPs) (End plate potentials or EPSPs) evoked by an impulse are increased when paired by a prior impulse. It is a form of short-term plasticity that enhances synaptic transmission for less than a second (Jackman and Regehr, 2017). When two stimuli are delivered within a short interval, the response to the second stimulus can be either enhanced or depressed relative to the response to the first stimulus (Zucker and Regehr, 2002; Katz and Miledi, 1968). Paired-pulse depression is commonly observed at all synapses at short (less than 20 ms) inter stimulus intervals, and most probably results from inactivation of voltage-dependent sodium or calcium channels or from a transient depletion of the readily releasable pool (RRP) of vesicles docked at the presynaptic terminal. A common scenario is that, in response to strong postsynaptic depolarization, dendrites release retrograde messengers that act through G-protein-coupled receptors located on presynaptic terminals to influence neurotransmitter release. Retrograde signaling by postsynaptic release of endocannabinoids can be initiated by strong depolarization or activation of postsynaptic metabotropic receptors and has been shown to transiently suppress inhibitory and excitatory synapses in several brain regions (Chevaleyre et al., 2006).

Facilitation helps counter the natural tendency of synapses to get weaken during repeated activation. At synapses with a low initial release probability, closely spaced presynaptic action potentials can result in facilitation, a short-term enhancement in which each subsequent action potential evokes greater neurotransmitter release (Zucker and Regehr, 2002). Synaptic transmission was first studied at the frog neuromuscular junction (NMJ), where one of the initial observations was that synaptic responses become much larger when presynaptic cells are activated by closely spaced stimuli (Eccles et al., 1941; Feng, 1940). Facilitation enhances readily releasable pool (RRP) of synaptic vesicles and neurotransmitter release considerably and can profoundly influence information transfer across synapses (Abbott and Regehr, 2004). Action-potential-evoked increases in presynaptic Ca^{2+} consists of a large, brief localized Ca^{2+} signal that activates the low-affinity Ca^{2+} sensor synaptotagmin 1 to trigger neurotransmitter release (Geppert et al., 1994), and a small residual Ca^{2+} signal that persists for tens of milliseconds, has been implicated in facilitation (Zucker and Regehr, 2002). Furthermore, depletion of the RRP occurs when synapses are activated at high frequencies for prolonged periods, and consequently synapses exhibit frequency dependent depression. Nonetheless, at some synapses, facilitation counteracts depression by increasing the probability of release (p) in a frequency-dependent manner (MacLeod et al., 2007). The same synapse can display either facilitation or depression, depending on its recent history of activation and modulation. Synapses that begin with a very high probability of transmitter release tend to depress their response to the second pulse (Dobrunz and Stevens, 1997). In contrast, synapses with a low initial probability of transmitter release normally exhibit an increase in p in response to a second stimulus.

11. Bi-Directional Synaptic Plasticity

A key concept is that, synaptic strength at excitatory synapses is bidirectionally modifiable by different patterns of activity. Embryonic striatal grafts create functional connections with the host striatal circuitry and restore synaptic transmission, facilitating the recovery of bidirectional synaptic plasticity (Mazzocchi-Jones et al., 2009). To test functional integration extracellular recordings was performed, the recording electrode was placed within the core of the graft region in a ventral location to the stimulating electrode. The graft-mediated expression of synaptic plasticity offers a neuronal substrate for the resumption of behavioural plasticity (Mayer et al., 1992; Brasted et al., 1999). The above observations demonstrate that, striatal grafts are anatomically connected and physiologically active. The transplanted neurons not only restore lost connections but also integrates as a functional component within a plastic neuronal system, necessary to learn and perform skilled actions.

12. Nociception Mediated Synaptic Plasticity

Tissue damage cause enormous emotional and psychological changes like depression and anxiety which significantly influence quality of life. The memory of past pain perception has been found to influence the subsequent experience of pain and plays a role in the prevention of potential future threats (Gedney and Logan, 2006; Noel et al., 2012). Genesis of pain is associated with enhanced neuronal activity (hyperalgesia) and sensitization. "Sensitizing soup" released following tissue damage recruit additional nociceptors and at the same time lower the threshold of activation, causing central and peripheral sensitization. Sensitization of nociceptors brings long-term synaptic plasticity in the spinal cord, subcortical areas and cortical areas contributing to the maintenance of chronic pain state (Zhuo, 2008; Zhuo, 2013). The anterior cingulate cortex (ACC) and other cortical areas, including the insular cortex, primary somatosensory cortex (S1), secondary somatosensory cortex (S2) and prefrontal cortex (PFC), are activated by various noxious stimuli (Talbot et al., 1991; Bushnell et al., 2013). This leads to the development of pain memory. At the synaptic level, it has been shown that the exposure to transient noxious heat does not cause alterations in either presynaptic LTP or postsynaptic LTP in the murine Anterior Cingulate Cortex (ACC), whereas neuropathic injury affects both forms of LTP, suggesting that enhanced transmission in the ACC is associated with chronic but not acute pain (Li et al, 2010; Koga et al., 2015). Rodent models of chronic inflammatory pain, neuropathic pain, bone-cancer pain and chronic visceral pain are all associated with the activation of Fos and Egr1 in ACC neurons (Wei et al., 1999; Drdla and Sandkühler, 2008). The induction of LTP in the ACC is required for the physiological processing of painful stimuli, including fear memory, whereas dysregulated or excessive LTP in the ACC might contribute to the chronic pain state.

FUTURE DIRECTIONS

The brain encodes external and internal events as complex, spatio-temporal patterns of activity in neurons that form 'neural circuits.' A key feature defining the behavior of any given neural circuit is the pattern of synaptic weights that connect the individual neurons consisting the circuit. Complex circuits are altered by experience to store new information and modify behavior. In addition, synaptic plasticity plays a key role in the early development of neural circuitry and maladaptive synaptic plasticity could impair behavioral functions contributing to neuropsychiatric disorders. Further, elucidation of the mechanisms of synaptic plasticity is needed to mark potential therapeutic targets for diseases in which brain plasticity is dysfunctional. High-frequency stimulation protocols and in vitro assays cannot be seen to resemble natural firing patterns or conditions found in the brain. More physiological experimental conditions, especially in vivo recording in awake animals, could lead the way to the development of improved models of learning mechanisms that could better correlate with the learning abilities of animals. By combining electrophysiological, molecular, cell biological and proteomic approaches, one could envision that the next few decades will bring further clarification on the detailed molecular mechanisms underlying the major forms of synaptic plasticity. Together, these approaches will allow sophisticated manipulation of synaptic plasticity mechanisms in highly restricted neural circuits and could greatly advance our understanding of how synaptic plasticity mediates both adaptive and pathological experience-dependent plasticity. Further, the development of methodologies for super-resolution time-lapse imaging of synapses, neurons and circuits in live animals promises to take the field forward and technique of specific synapse elimination using genetic approaches could make it possible to explain neural processing during normal and diseased.

REFERENCES

Abbott LF, Regehr WG (2004) Synaptic computation. *Nature.* 431(7010):796-803.

Abbott LF, Varela JA, Sen K, Nelson SB (1997) Synaptic depression and cortical gain control. *Science.* 275(5297):220-4.

Anggono V, Koç-Schmitz Y, Widagdo J, Kormann J, Quan A, Chen CM, Robinson PJ, Choi SY, Linden DJ, Plomann M, Huganir RL (2013) PICK1 interacts with PACSIN to regulate AMPA receptor internalization and cerebellar long-term depression. *Proc Natl Acad Sci US.* 110(34):13976-81.

Araque A, Carmignoto G, Haydon PG, Oliet SH, Robitaille R, Volterra A (2014) Gliotransmitters travel in time and space. *Neuron.* 81(4):728-39.

Atluri PP, Regehr WG (1996) Determinants of the time course of facilitation at the granule cell to Purkinje cell synapse. *J Neurosci.* 16(18):5661-71.

Bao J, Reim K, Sakaba T (2010) Target-dependent feedforward inhibition mediated by short-term synaptic plasticity in the cerebellum. *J Neurosci.* 30(24):8171-9.

Barria A, Muller D, Derkach V, Griffith LC, Soderling TR (1997) Regulatory phosphorylation of AMPA-type glutamate receptors by CaM-KII during long-term potentiation. *Science.* 276(5321):2042-5.

Bartley AF, Dobrunz LE (2015) Short-term plasticity regulates the excitation/inhibition ratio and the temporal window for spike integration in CA1 pyramidal cells. *Eur J Neurosci.* 41(11):1402-15.

Bartley AF, Huang ZJ, Huber KM, Gibson JR (2008) Differential activity-dependent, homeostatic plasticity of two neocortical inhibitory circuits. *J Neurophysiol.* 100(4):1983-94.

Bartos M, Elgueta C (2012) Functional characteristics of parvalbumin- and cholecystokinin-expressing basket cells. *J Physiol.* 590(4):669-81.

Berger TW (1984) Long-term potentiation of hippocampal synaptic transmission affects rate of behavioral learning. *Science.* 224(4649):627-30.

Berger TW, Alger B, Thompson RF (1976) Neuronal substrate of classical conditioning in the hippocampus. *Science*. 192(4238):483-5.

Berger TW, Thompson RF (1977) Limbic system interrelations: functional division among hippocampal-septal connections. *Science*. 197(4303):587-9.

Bliss TV, Collingridge GL (1993) A synaptic model of memory: long-term potentiation in the hippocampus. *Nature*. 361(6407):31-9.

Bliss TV, Lomo T (1973) Long-lasting potentiation of synaptic transmission in the dentate area of the anaesthetized rabbit following stimulation of the performant path. *J Physiol*. 232(2):331-56.

Bloch V (1970) Facts and hypotheses concerning memory consolidation processes. *Brain Res*. 24(3):561-75.

Bloch V and Laroche S (1984) Facts and hypotheses related to the search for the engram. In: *Neurobiology of Learning and Memory,* pp. 249-260. Eds. G. Lynch, J. L. McGaugh and N. M. Weinberger. Guildford Press, New York.

Bloch V, Laroche S (1981) Conditioning of hippocampal cells: its acceleration and long-term facilitation by post-trial reticular stimulation. *Behav Brain Res*. 3(1):23-42.

Braitenberg V and Schüz A (1999) Cortex: Statistics and Geometry of Neuronal Connectivity, 2nd edn. *J Anat*. 194(Pt 1): 153–157.

Bramham CR, Maho C, Laroche S (1994) Suppression of long-term potentiation induction during alert wakefulness but not during 'enhanced' REM sleep after avoidance learning. *Neuroscience*. 59(3):501-9.

Brasted PJ, Watts C, Robbins TW, Dunnett SB (1999) Associative plasticity in striatal transplants. *Proc Natl Acad Sci US*. 96(18): 10524-9.

Burrone J, O'Byrne M, Murthy VN (2002) Multiple forms of synaptic plasticity triggered by selective suppression of activity in individual neurons. *Nature*. 420(6914):414-8.

Bushnell MC, Ceko M, Low LA (2013) Cognitive and emotional control of pain and its disruption in chronic pain. *Nat Rev Neurosci*. 14(7):502-11.

Buzsàki G, Grastyàn E, Czopf J, Kellènyi L, Prohaska O (1981) Changes in neuronal transmission in the rat hippocampus during behavior. *Brain Res*. 225(2):235-47.

Cajal, RS (1894) La Fine Structure des Centres Nerveux. *Proc. R. Soc. Lond.* 55 331-335 444-468.

Calabresi P, Pisani A, Mercuri NB, Bernardi G (1992) Long-term Potentiation in the Striatum is Unmasked by Removing the Voltage-dependent Magnesium Block of NMDA Receptor Channels. *Eur J Neurosci.* 4(10):929-935.

Calabresi P, Pisani A, Rothwell J, Ghiglieri V, Obeso JA, Picconi B (2016) Hyperkinetic disorders and loss of synaptic downscaling. *Nat Neurosci.* 19(7):868-75.

Castellucci V, Pinsker H, Kupfermann I, Kandel ER (1970) Neuronal mechanisms of habituation and dishabituation of the gill-withdrawal reflex in Aplysia. *Science.* 167(3926):1745-8.

Castro CA, Silbert LH, McNaughton BL, Barnes CA (1989) Recovery of spatial learning deficits after decay of electrically induced synaptic enhancement in the hippocampus. *Nature.* 342(6249):545-8.

Chaillan FA, Roman FS, Soumireu-Mourat B (1998) Apprentissage olfactif: les systeÁmes de meÂmoire chez le rat. *C. R. Acad. Sci.* (Paris) 321, 157-161.

Charpier S, Deniau JM (1997) In vivo activity-dependent plasticity at cortico-striatal connections: evidence for physiological long-term potentiation. *Proc Natl Acad Sci US.* 94(13):7036-40.

Cheetham CE, Park U, Belluscio L (2016) Rapid and continuous activity-dependent plasticity of olfactory sensory input. *Nat Commun.* 7:10729.

Chevaleyre V, Takahashi KA, Castillo PE (2006) Endocannabinoid-mediated synaptic plasticity in the CNS. *Annu Rev Neurosci.* 29:37-76.

Chung HJ, Steinberg JP, Huganir RL, Linden DJ (2003) Requirement of AMPA receptor GluR2 phosphorylation for cerebellar long-term depression. *Science.* 300(5626):1751-5.

Coesmans M, Weber JT, De Zeeuw CI, Hansel C (2004) Bidirectional parallel fiber plasticity in the cerebellum under climbing fiber control. *Neuron.* 44(4):691-700.

Croll SD, Sharp PE, Bostock E (1992) Evidence for NMDA receptor involvement in environmentally induced dentate gyrus plasticity. *Hippocampus*. 2(1):23-8.

Davies CH, Davies SN, Collingridge GL (1990) Paired-pulse depression of monosynaptic GABA-mediated inhibitory postsynaptic responses in rat hippocampus. *J Physiol*. 424:513-31.

Desai NS, Cudmore RH, Nelson SB, Turrigiano GG (2002) Critical periods for experience-dependent synaptic scaling in visual cortex. *Nat Neurosci*. 5(8):783-9.

Ding J, Peterson JD, Surmeier DJ (2008) Corticostriatal and thalamostriatal synapses have distinctive properties. *J Neurosci*. 28(25):6483-92.

Dittman JS, Kreitzer AC, Regehr WG (2000) Interplay between facilitation, depression, and residual calcium at three presynaptic terminals. *J Neurosci*. 20(4):1374-85.

Dobrunz LE, Stevens CF (1997) Heterogeneity of release probability, facilitation, and depletion at central synapses. *Neuron*. 18(6):995-1008.

Doyère V, Burette F, Negro CR, Laroche S (1993) Long-term potentiation of hippocampal afferents and efferents to prefrontal cortex: implications for associative learning. *Neuropsychologia*. 31(10):1031-53

Drdla R, Sandkühler J (2008) Long-term potentiation at C-fibre synapses by low-level presynaptic activity in vivo. *Mol Pain*. 4:18.

Dudai Y, (1988) Neurogenetic dissection of learning and short-term memory in Drosophila. *Annu Rev Neurosci*.;11:537-63.

Eccles JC, Katz B, Kuffler SW (1941) Nature of the “endplate potential” in curarized muscle. *J. Neurophysiol*. 4, 362–387.

Eichenbaum H (2004) Hippocampus: cognitive processes and neural representations that underlie declarative memory. *Neuron*. 44(1):109-20.

Feng TP (1940) Studies on the neuromuscular junction. XVIII. The local potentials around n-m junctions induced by single and multiple volleys. *Chin. J. Physiol*. 15, 367–404.

Fino E, Glowinski J, Venance L (2005) Bidirectional activity-dependent plasticity at corticostriatal synapses. *J Neurosci.* 25(49): 11279-87.

Fukunaga K, Muller D, Miyamoto E (1995) Increased phosphorylation of Ca2+/calmodulin-dependent protein kinase II and its endogenous substrates in the induction of long-term potentiation. *J Biol Chem.* 270(11):6119-24.

Gedney JJ, Logan H (2006) Pain related recall predicts future pain report. Pain. 121(1-2):69-76.

Geppert M, Goda Y, Hammer RE, Li C, Rosahl TW, Stevens CF, Südhof TC (1994) Synaptotagmin I: a major Ca2+ sensor for transmitter release at a central synapse. *Cell.* 79(4):717-27.

Gerdeman GL, Ronesi J, Lovinger DM (2002) Postsynaptic endocannabinoid release is critical to long-term depression in the striatum. *Nat Neurosci*. 5(5):446-51.

Gerrow K, Triller A (2010) Synaptic stability and plasticity in a floating world. *Curr Opin Neurobiol*. 20(5):631-9.

Giese KP, Fedorov NB, Filipkowski RK, Silva AJ (1998) Autophosphorylation at Thr286 of the alpha calcium-calmodulin kinase II in LTP and learning. *Science.* 279(5352):870-3.

Goel A, Jiang B, Xu LW, Song L, Kirkwood A, Lee HK (2006) Cross-modal regulation of synaptic AMPA receptors in primary sensory cortices by visual experience. *Nat Neurosci*. 9(8):1001-3.

Green EJ, McNaughton BL, Barnes CA (1990) Exploration-dependent modulation of evoked responses in fascia dentata: dissociation of motor, EEG, and sensory factors and evidence for a synaptic efficacy change. *J Neurosci*. 10(5):1455-71.

Harnett MT, Makara JK, Spruston N, Kath WL, Magee JC (2012) Synaptic amplification by dendritic spines enhances input cooperativity. *Nature.* 491(7425):599-602.

Hars B, Hennevin E, Pasques P (1985) Improvement of learning by cueing during postlearning paradoxical sleep. *Behav Brain Res*. 18(3):241-50.

Haydon PG (2001) GLIA: listening and talking to the synapse. *Nat Rev Neurosci.* 2(3):185-93.

Hebb DO (2005) *The Organization of Behavior: A Neuropsychological Theory*. New York: Psychology Press.

Henley JM, Wilkinson KA (2013) AMPA receptor trafficking and the mechanisms underlying synaptic plasticity and cognitive aging. *Dialogues Clin Neurosci*. 15(1):11-27.

Hennevin E, Maho C, Hars B, Dutrieux G (1993) Learning-induced plasticity in the medial geniculate nucleus is expressed during paradoxical sleep. *Behav Neurosci*. 107(6):1018-30.

Hernandez A, Tan C, Mettlach G, Pozo K, Plattner F, Bibb JA (2016) Cdk5 Modulates Long-Term Synaptic Plasticity and Motor Learning in Dorsolateral Striatum. *Sci Rep*. 6:29812.

Hernandez AI, Blace N, Crary JF, Serrano PA, Leitges M, Libien JM, Weinstein G, Tcherapanov A, Sacktor TC (2003) Protein kinase M zeta synthesis from a brain mRNA encoding an independent protein kinase C zeta catalytic domain. Implications for the molecular mechanism of memory. *J Biol Chem*. 278(41):40305-16.

Holtmaat A, Svoboda K (2009) Experience-dependent structural synaptic plasticity in the mammalian brain. *Nat Rev Neurosci.* 10(9):647-58.

Ito M, Sakurai M, Tongroach P (1982) Climbing fibre induced depression of both mossy fibre responsiveness and glutamate sensitivity of cerebellar Purkinje cells. *The Journal of Physiology*. 324:113-134.

Ito, M (1989) Long-term depression. A. *Rev. Neurosci*. 12, 85-102.

Jackman SL, Regehr WG (2017) The Mechanisms and Functions of Synaptic Facilitation. *Neuron.* 94(3):447-464.

Jörntell H, Hansel C (2006) Synaptic memories upside down: bidirectional plasticity at cerebellar parallel fiber-Purkinje cell synapses. *Neuron*. 52(2):227-38.

Kalia LV, Gingrich JR, Salter MW (2004) Src in synaptic transmission and plasticity. *Oncogene*. 23(48):8007-16.

Kandel ER (2001) The molecular biology of memory storage: a dialogue between genes and synapses. *Science*. 294(5544):1030-8.

Kandel ER, Schwartz JH (1982) Molecular biology of learning: modulation of transmitter release. *Science*. 218(4571):433-43.

Katz B, Miledi R (1968) The role of calcium in neuromuscular facilitation. *J Physiol*. 195(2):481-92.

Kelly MT, Yao Y, Sondhi R, Sacktor TC (2007) Actin polymerization regulates the synthesis of PKMzeta in LTP. *Neuropharmacology*. 52(1):41-5.

Kirkwood A, Dudek SM, Gold JT, Aizenman CD, Bear MF (1993) Common forms of synaptic plasticity in the hippocampus and neocortex in vitro. *Science*. 260(5113):1518-21.

Kispersky TJ, Fernandez FR, Economo MN, White JA (2012) Spike resonance properties in hippocampal O-LM cells are dependent on refractory dynamics. *J Neurosci*. 32(11):3637-51.

Koga K, Descalzi G, Chen T, Ko HG, Lu J, Li S, Son J, Kim T, Kwak C, Huganir RL, Zhao MG, Kaang BK, Collingridge GL, Zhuo M (2015) Coexistence of two forms of LTP in ACC provides a synaptic mechanism for the interactions between anxiety and chronic pain. *Neuron*. 85(2):377-89.

Kuhn M, Wolf E, Maier JG, Mainberger F, Feige B, Schmid H, Bürklin J, Maywald S, Mall V, Jung NH, Reis J, Spiegelhalder K, Klöppel S, Sterr A, Eckert A, Riemann D, Normann C, Nissen C (2016) Sleep recalibrates homeostatic and associative synaptic plasticity in the human cortex. *Nat Commun*. 7:12455.

Laroche S, DoyeÁ re V, ReÂdini-Del Negro C (1991) What role for long-term potentiation in learning and the maintenance of memories? In: Long-term Potentiation: *A Debate of Current Issues*, pp. 301-316. Eds. M. Baudry and J. L. Davis. MIT Press, Cambridge.

Laroche S, Errington ML, Lynch MA, Bliss TV (1987) Increase in [3H]glutamate release from slices of dentate gyrus and hippocampus following classical conditioning in the rat. *Behav Brain Res*. 25(1):23-9.

Laroche S, Falcou R, Bloch V (1983) Post-trial reticular facilitation of associative changes in multiunit activity; comparison between dentate gyrus and entorhinal cortex. *Behav Brain Res*. 9(3):381-7.

Li XY, Ko HG, Chen T, Descalzi G, Koga K, Wang H, Kim SS, Shang Y, Kwak C, Park SW, Shim J, Lee K, Collingridge GL, Kaang BK, Zhuo

M (2010) Alleviating neuropathic pain hypersensitivity by inhibiting PKMzeta in the anterior cingulate cortex. *Science.* 330(6009):1400-4.

Lisman JE (1997) Bursts as a unit of neural information: making unreliable synapses reliable. *Trends Neurosci*. 20(1):38-43.

LoTurco JL, Coulter DA, Alkon DL (1988) Enhancement of synaptic potentials in rabbit CA1 pyramidal neurons following classical conditioning. *Proc Natl Acad Sci US*. 85(5):1672-6.

MacLeod KM, Horiuchi TK, Carr CE (2007) A role for short-term synaptic facilitation and depression in the processing of intensity information in the auditory brain stem. *J Neurophysiol*. 97(4):2863-74.

Magee JC, Johnston D (1997) A synaptically controlled, associative signal for Hebbian plasticity in hippocampal neurons. *Science.* 275(5297):209-13.

Major G, Larkum ME, Schiller J (2013) Active properties of neocortical pyramidal neuron dendrites. *Annu Rev Neurosci*. 36:1-24.

Malenka RC, Bear MF (2004) LTP and LTD: an embarrassment of riches. *Neuron*. 44(1):5-21. Review. PubMed PMID: 15450156.

Malenka RC, Kauer JA, Zucker RS, Nicoll RA (1988) Postsynaptic calcium is sufficient for potentiation of hippocampal synaptic transmission. *Science*. 242(4875):81-4.

Maren S (2001) Neurobiology of Pavlovian fear conditioning. *Annu Rev Neurosci*. 24:897-931.

Maren S (2005) Synaptic mechanisms of associative memory in the amygdala. *Neuron.* 7(6):783-6.

Matsuzaki M, Honkura N, Ellis-Davies GC, Kasai H (2004) Structural basis of long-term potentiation in single dendritic spines. *Nature*. 429(6993):761-6.

Mayer E, Brown VJ, Dunnett SB, Robbins TW (1992) Striatal graft-associated recovery of a lesion-induced performance deficit in the rat requires learning to use the transplant. *Eur J Neurosci*. 4(2):119-26.

Mazzocchi-Jones D, Döbrössy M, Dunnett SB (2009) Embryonic striatal grafts restore bi-directional synaptic plasticity in a rodent model of Huntington's disease. *Eur J Neurosci*. 30(11):2134-42.

McNamara RK, Kirkby RD, dePape GE, Skelton RW, Corcoran ME (1993) Differential effects of kindling and kindled seizures on place learning in the Morris water maze. *Hippocampus.* 3(2):149-52.

Migues PV, Hardt O, Wu DC, Gamache K, Sacktor TC, Wang YT, Nader K (2010) PKMzeta maintains memories by regulating GluR2-dependent AMPA receptor trafficking. *Nat Neurosci.* 13(5):630-4.

Moser EI, Krobert KA, Moser MB, Morris RG (1998) Impaired spatial learning after saturation of long-term potentiation. *Science.* 281(5385):2038-42.

Muslimov IA, Nimmrich V, Hernandez AI, Tcherepanov A, Sacktor TC, Tiedge H (2004) Dendritic transport and localization of protein kinase Mzeta mRNA: implications for molecular memory consolidation. *J Biol Chem.* 279(50):52613-22.

Neuenschwander-El Massioui N, Dutrieux G, Edeline J.-M (1991) Conditioned hippocampal cellular response to a behaviorally silent conditioned stimulus. *Behav. Neurosci.* 105, 91-103.

Neveu D, Zucker RS (1996) Postsynaptic levels of [Ca2+]i needed to trigger LTD and LTP. *Neuron.* 16(3):619-29.

Noel M, Chambers CT, McGrath PJ, Klein RM, Stewart SH (2012) The influence of children's pain memories on subsequent pain experience. *Pain.* 153(8):1563-72.

Okabe S, Kim HD, Miwa A, Kuriu T, Okado H (1999) Continual remodeling of postsynaptic density and its regulation by synaptic activity. *Nat Neurosci.* 2(9):804-11.

Paradis S, Sweeney ST, Davis GW (2001) Homeostatic control of presynaptic release is triggered by postsynaptic membrane depolarization. *Neuron.* 30(3):737-49. Erratum in: *Neuron* 2001 Jul 19;31(1):167.

Partridge JG, Tang KC, Lovinger DM (2000) Regional and postnatal heterogeneity of activity-dependent long-term changes in synaptic efficacy in the dorsal striatum. *J Neurophysiol.* 84(3):1422-9.

Ramirez OA and Carrer HF (1989) Correlation between threshold to induce long-term potentiation in the hippocampus and performance in a shuttle box avoidance response in rats. *Neurosci Lett.* 104(1-2):152-6.

Ramírez OA, Orsingher OA, Carrer HF (1988) Differential threshold for long-term potentiation in the hippocampus of rats with inborn high or low learning capacity. *Neurosci Lett*. 92(3):275-9.

Reymann KG, Frey JU (2007) The late maintenance of hippocampal LTP: requirements, phases, 'synaptic tagging,' 'late-associativity' and implications. *Neuropharmacology*. 52(1):24-40.

Rinzel J, Rall W (1974) Transient response in a dendritic neuron model for current injected at one branch. *Biophys J*. 14(10):759-90.

Rioult-Pedotti MS, Friedman D, Donoghue JP (2000) Learning-induced LTP in neocortex. *Science*. 290(5491):533-6.

Rioult-Pedotti MS, Friedman D, Hess G, Donoghue JP (1998) Strengthening of horizontal cortical connections following skill learning. *Nat Neurosci*. 1(3):230-4.

Rodrigues SM, Schafe GE, LeDoux JE (2004) Molecular mechanisms underlying emotional learning and memory in the lateral amygdala. *Neuron*. 44(1):75-91.

Roman F, Staubli U, Lynch G (1987) Evidence for synaptic potentiation in a cortical network during learning. *Brain Res*. 418, 221-226.

Rose GJ, Fortune ES (1999) Frequency-dependent PSP depression contributes to low-pass temporal filtering in Eigenmannia. *J Neurosci*. 19(17):7629-39.

Rozov A, Burnashev N, Sakmann B, Neher E (2001) Transmitter release modulation by intracellular Ca2+ buffers in facilitating and depressing nerve terminals of pyramidal cells in layer 2/3 of the rat neocortex indicates a target cell-specific difference in presynaptic calcium dynamics. *J Physiol*. 531(Pt 3):807-26.

Ruediger S, Spirig D, Donato F, Caroni P (2012) Goal-oriented searching mediated by ventral hippocampus early in trial-and-error learning. *Nat Neurosci*. 2012 Nov; 15(11):1563-71.

Ruèthrich H, Matthies H, Ott T (1982) Long-term changes in synaptic excitability of hippocampal cell populations as a result of training. In: *Neuronal Plasticity and Memory Formation*, pp. 589-594. Eds. C. A. Marsan and H. Matthies. Raven Press, New York.

Rumpel S, LeDoux J, Zador A, Malinow R (2005) Postsynaptic receptor trafficking underlying a form of associative learning. *Science.* 308(5718):83-8.

Sacktor TC (2008) PKMzeta LTP maintenance, and the dynamic molecular biology of memory storage. *Prog Brain Res.* 169:27-40.

Sale MV, Ridding MC, Nordstrom MA (2008) Cortisol inhibits neuroplasticity induction in human motor cortex. *J Neurosci.* 28(33):8285-93.

Sanes JR, Lichtman JW (1999) Can molecules explain long-term potentiation? *Nat Neurosci.* 2(7):597-604.

Schiller J, Schiller Y, Clapham DE (1998) NMDA receptors amplify calcium influx into dendritic spines during associative pre- and postsynaptic activation. *Nat Neurosci.* 1(2):114-8.

Shema R, Haramati S, Ron S, Hazvi S, Chen A, Sacktor TC, Dudai Y (2011) Enhancement of consolidated long-term memory by overexpression of protein kinase Mzeta in the neocortex. *Science.* 331(6021):1207-10.

Shema R, Sacktor TC, Dudai Y (2007) Rapid erasure of long-term memory associations in the cortex by an inhibitor of PKM zeta. *Science.* 317(5840):951-3.

Shen W, Flajolet M, Greengard P, Surmeier DJ (2008) Dichotomous dopaminergic control of striatal synaptic plasticity. *Science.* 321(5890):848-51.

Silva AJ, Paylor R, Wehner JM, Tonegawa S (1992) Impaired spatial learning in alpha-calcium-calmodulin kinase II mutant mice. *Science.* 257(5067):206-11.

Sjöström PJ, Rancz EA, Roth A, Häusser M (2008) Dendritic Excitability and Synaptic Plasticity. *Physiological Reviews* 88 (2), 769-840.

Sjöström PJ, Turrigiano GG, Nelson SB (2001) Rate, timing, and cooperativity jointly determine cortical synaptic plasticity. *Neuron.* 32(6):1149-64.

Skelton RW, Scarth AS, Wilkie DM, Miller JJ, Phillips AG (1987) Long-term increase in dentate granule cell responsivity accompany operant conditioning. *J Neurosci.* 7(10):3081-7.

Spencer JP, Murphy KP (2000) Bi-directional changes in synaptic plasticity induced at corticostriatal synapses in vitro. *Exp Brain Res.* 135(4):497-503.

Squire LR and Zola SM (1996) Structure and function of declarative and nondeclarative memory systems. *Proc. Natl Acad. Sci. US* 93, 13515-13522.

Stent GS (1973) A physiological mechanism for Hebb's postulate of learning. *Proc Natl Acad Sci U S A.* 70(4):997-1001.

Stevens CF, Wesseling JF (1999) Augmentation is a potentiation of the exocytotic process. *Neuron.* 22(1):139-46.

Sun HY, Lyons SA, Dobrunz LE (2005) Mechanisms of target-cell specific short-term plasticity at Schaffer collateral synapses onto interneurones versus pyramidal cells in juvenile rats. *J Physiol.* 568(Pt 3):815-40.

Sweatt JD (2004) Mitogen-activated protein kinases in synaptic plasticity and memory. *Curr Opin Neurobiol.* 14(3):311-7.

Tadavarty R, Kaan TK, Sastry BR (2009) Long-term depression of excitatory synaptic transmission in rat hippocampal CA1 neurons following sleep-deprivation. *Exp Neurol.* 216(1):239-42.

Talbot JD, Marrett S, Evans AC, Meyer E, Bushnell MC, Duncan GH (1991) Multiple representations of pain in human cerebral cortex. *Science.* 251(4999):1355-8.

Tanaka K, Augustine GJ (2008) A positive feedback signal transduction loop determines timing of cerebellar long-term depression. *Neuron.* 59(4):608-20.

Thiagarajan TC, Lindskog M, Tsien RW (2005) Adaptation to synaptic inactivity in hippocampal neurons. *Neuron.* 47(5):725-37.

Thomas GM, Huganir RL (2004) MAPK cascade signalling and synaptic plasticity. *Nat Rev Neurosci.* 5(3):173-83.

Thompson RF (1990) Neural mechanisms of classical conditioning in mammals. *Phil. Trans. R. Soc. Lond. [Biol.]* 329, 161-170.

Turrigiano GG, Leslie KR, Desai NS, Rutherford LC, Nelson SB (1998) Activity-dependent scaling of quantal amplitude in neocortical neurons. *Nature.* 391(6670):892-6.

Turrigiano GG, Nelson SB (2004) Homeostatic plasticity in the developing nervous system. *Nat Rev Neurosci.* 5(2):97-107.

Tye KM, Tye LD, Cone JJ, Hekkelman EF, Janak PH, Bonci A (2010) Methylphenidate facilitates learning-induced amygdala plasticity. *Nat Neurosci.* 13(4):475-81.

Valtcheva S, Venance L (2016) Astrocytes gate Hebbian synaptic plasticity in the striatum. *Nat Commun.* 7:13845.

Vyazovskiy VV, Cirelli C, Pfister-Genskow M, Faraguna U, Tononi G (2008) Molecular and electrophysiological evidence for net synaptic potentiation in wake and depression in sleep. *Nat Neurosci.* 11(2):200-8.

Weber JP, Andrásfalvy BK, Polito M, Magó Á, Ujfalussy BB, Makara JK (2016) Location-dependent synaptic plasticity rules by dendritic spine cooperativity. *Nat Commun.* 7:11380.

Wei F, Li P, Zhuo M (1999) Loss of synaptic depression in mammalian anterior cingulate cortex after amputation. *J Neurosci.* 19(21):9346-54.

Weisz D, Clark GA, Thompson RF (1984) Increased responsivity of dentate granule cells during nictitating membrane response conditioning in rabbit. *Behav. Brain Res.* 12, 145-154.

Whitlock JR, Heynen AJ, Shuler MG, Bear MF (2006) Learning induces long-term potentiation in the hippocampus. *Science.* 313(5790):1093-7.

Wierenga CJ, Ibata K, Turrigiano GG (2005) Postsynaptic expression of homeostatic plasticity at neocortical synapses. *J Neurosci.* 25(11):2895-905.

Wierenga CJ, Wadman WJ (2003) Excitatory inputs to CA1 interneurons show selective synaptic dynamics. *J Neurophysiol.* 90(2):811-21.

Wierenga CJ, Walsh MF, Turrigiano GG (2006) Temporal regulation of the expression locus of homeostatic plasticity. *J Neurophysiol.* 96(4):2127-33.

Xing B, Li YC, Gao WJ (2016) GSK3β Hyperactivity during an Early Critical Period Impairs Prefrontal Synaptic Plasticity and Induces Lasting Deficits in Spine Morphology and Working Memory. *Neuropsychopharmacology.* 41(13):3003-3015.

Xu C, Zhao MX, Poo MM, Zhang XH (2008) GABA(B) receptor activation mediates frequency-dependent plasticity of developing GABAergic synapses. *Nat Neurosci*. 11(12):1410-8.

Yang Y, Wang ZH, Jin S, Gao D, Liu N, Chen SP, Zhang S, Liu Q, Liu E, Wang X, Liang X, Wei P, Li X, Li Y, Yue C, Li HL, Wang YL, Wang Q, Ke D, Xie Q, Xu F, Wang L, Wang JZ (2016) Opposite monosynaptic scaling of BLP-vCA1 inputs governs hopefulness- and helplessness-modulated spatial learning and memory. *Nat Commun.* 7:11935.

Zhuo M (2008) Cortical excitation and chronic pain. *Trends Neurosci.* 31(4):199-207.

Zhuo M (2013) Long-term potentiation in the anterior cingulate cortex and chronic pain. *Philos Trans R Soc Lond B Biol Sci.* 369(1633):20130146.

Zola-Morgan SM and Squire LR (1990) The primate hippocampal formation: evidence for a time-limited role in memory storage. *Science* 250, 288-290.

Zucker RS, Regehr WG (2002) Short-term synaptic plasticity. Annu *Rev Physiol*. 64:355-405.

In: Synaptic Plasticity
Editor: Sraboni Chaudhury
ISBN: 978-1-53613-606-7

Chapter 5

SYNAPTIC PLASTICITY IN AUTISM SPECTRUM DISORDERS

***Meghna Saxena*[1] *and Kapil Saxena*[2,3]**
[1]Department of Neurosurgery, University of Michigan,
Ann Arbor, MI, US
[2]The Patrick Wild Centre, University of Edinburgh, UK
[3]Center for Brain Development and Repair,
Institute of Stem Cell Biology and Regenerative Medicine,
Bangalore, India

ABSTRACT

Autism is one of the most common neurodevelopmental disorders and has a high impact on families and societies worldwide. According to the Centers for Disease Control and Prevention (CDC), one in 68 children are currently diagnosed with some form of autism. Because there is a broad range of symptoms across patients, with varying levels of severity, autism is described as a spectrum disorder. Patients suffering from autism spectrum disorders have severe learning difficulties, with prominent repetitive behavior and inability to socialize normally. Autism is a lifelong condition requiring lifelong care; this affects not only immediate carers such as parents and relatives, but also the wider society. While the carers go through mental, physical and financial turmoil, the burden on society is majorly financial. In this chapter, we will focus on learning and memory aspects of autism spectrum disorder, paying special attention to synaptic plasticity. We will present an overall understanding based on studies involving animal models of autism spectrum disorders and will also cite some important human studies.

Keywords: autism, neurodevelopment, synaptic plasticity

1. Introduction

Autism is a neurodevelopmental disorder characterized by limited social communication skills, stereotyped and repetitive behaviors and abnormal sensorial sensitivity. Children with autism are usually diagnosed when they show a delay in the expression of standard neuro-developmental milestones and they are often intellectually disabled. The term *autism* was coined by Leo Kanner, an Austrian-American psychiatrist, who over half a century ago described characteristics of 11 children who had social and emotional difficulties (Kanner 1943). It was later noted that several disorders like Asperger's syndrome, childhood disintegrative disorder and pervasive developmental disorders-not otherwise specified (PDD-NOS) shared many common symptoms that had resulted in inconsistent diagnoses. This problem was addressed in 2013 in the Diagnostic and Statistical manual of Mental Disorders-5 (DSM-5), where a new umbrella term autism spectrum disorders (ASD) was introduced (APA 2013). ASD not only covers autistic disorders but also include Asperger's syndrome, childhood disintegrative disorder, Rett syndrome and PDD-NOS (APA 2013).

According to data from the CDC in March 2014 for the United States, one in 68 children has some form of autism whereas boys were more susceptible (one in 42 boys vs. one in 189 girls) (Gillberg 2010).

Research has discovered genes and gene mutations that are associated with autism in humans (Blumberg, Schieve, and Lu 2013). Autism can be caused by the mutation of a single gene, such as the FRM1 gene, where mutation leads to fragile x mental retardation. In addition, certain risk genes, when combined with environmental factors can trigger autism-like phenotypes in the developing brain. There is a plethora of environmental risk factors associated with autism. Modabberian et al. have recently published a meta-analysis of evidence-based human studies (Modabbernia, Velthorst, and Reichenberg 2017). They found that events before, during and after birth are of considerable importance, including advanced age of either parent at the time of conception. Other risk factors include maternal illness during pregnancy and certain birthing difficulties that specifically involve oxygen deprivation in the baby's brain. There is considerable evidence that suggests an association between heavy metals, such as mercury and lead with ASD. These factors on their own do not cause autism per se, but in combination with genetic risk factors, they appear to modestly increase autism risk. Mechanisms of how environmental risk factors lead to ASD are still a topic of debate and studies attempting to investigate these are still in their infancy. It is important for social awareness relating to autism that vaccinations, maternal smoking, thimerosal exposure and likely, assisted reproductive technology, are not related to ASD risk (Modabbernia, Velthorst, and Reichenberg 2017).

Synaptic plasticity is one process that may be involved in how environmental risk factors and their interaction with susceptible genes lead to ASD. Synaptic plasticity is the phenomenon where synapses change (e.g., strengthen or weaken) over time. The reason for these changes in synaptic strength is typically due to synapses being active or inactive and the structural component of synaptic plasticity is synapses (Bailey, Kandel, and Harris 2015). Many animal model studies investigating have associated ASD with synaptic plasticity, both in terms of synaptic strength or number of synapses involved (Sidorov, Auerbach, and Bear 2013, Sudhof 2008). In addition, there are many risk genes that are regulated by neural activity and seem to have role in synaptic plasticity. This has led to the hypothesis that mutations can lead to loss or gain of gene function, leading to an imbalance in general neural functioning, which then leads to an increased risk of ASD (Ung et al. 2017). In this chapter we will review the role of genetics in ASD and focus on the effects of synaptic plasticity on ASD.

2. Genetics and Epigenetics of ASD

Before we discuss the genetics associated with ASD it may be useful to remind ourselves of the diversity of the human genome. The average human has 3 million genetic variants stemming from a typical genome. These variants are mostly single nucleotide polymorphism (SNPs) or copy number variants (CNVs). Of these variants, many are common, but approximately 4% are rare, which amounts to a staggering1.3 million variants. Therefore, the possibility of carrying a disease related gene is extremely high and thus carrying a disease mutation is not the exception but the rule. This leads us to perhaps the two most important questions for epidemiologists and molecular geneticists regarding autism: how are mutations associated with varying degrees of autism risk distributed within the population and why is it that most individuals remain unaffected despite carrying deleterious genes? We know that a single highly penetrant gene mutation can cause ASD and also that an accumulation of thousands of low risk alleles can lead to a very similar phenotype, which suggests that ASD has a heterogenous genetic architecture.

The causal relationships between the contributions of genetics, epigenetics and environmental risk factors towards the risk of ASD are not clear, but all the epidemiological studies to date point to the fact that there is a strong genetic contribution to the risk of ASD. Data obtained from twin studies strengthen this claim: in families with a history of ASD, the chances of having a child with ASD increases due to a genetic contribution from the ASD parent and decreases with genetic distance from the individual with ASD (Ronald and Hoekstra 2011, Sandin et al. 2014). Furthermore, a large proportion of our knowledge and understanding of ASD genetics comes from studies on rare and highly penetrant variants from *de novo* clinical cases. This genetic understanding helped in the evolution of a new methodological approach to investigating ASD that is not limited to single gene causality but includes a neuronal signalling pathway approach (An and Claudianos 2016, Betancur 2011).

One of the biggest disappointments came when one of the largest genome wide association studies (GWASs), that included data from more than 5000 families with ASD, were not able to pinpoint to a single SNP that was associated with ASD (Anney et al. 2012, Cross-Disorder Group of the Psychiatric Genomics 2013). Despite this, the importance of SNPs should not be neglected as an estimated 17-60% of ASD cases can only be explained by the presence of SNPs (Cross-Disorder Group of the Psychiatric Genomics 2013).

Even with today's modern technology, the genetic causes of ASD can only be detected in 25% of cases; of these, 3-6% are from monogenic recessive forms 10-20% are from *de novo* deleterious mutations. It is the hope that the number of detections of ASD can be increased using whole genome sequencing as currently non-coding regions are not included as part of the regular screening process.

3. Synaptic Plasticity in ASD

The term *synaptic plasticity* refers to the phenomenon whereby synaptic activity leads to synapses making new connections or strengthening or weakening existing connections between two synapses.

Synaptic plasticity can be modulated by sensory inputs and intrinsic brain activity, which can cause long-term changes in synaptic efficacy that eventually results in a modulation of the number of synapses; this modulation ultimately affects neuronal connectivity. Many mutations have been identified that are associated with ASD and these mutations influence the basic structure and number of synapses. In addition, the mutated genes encode proteins that affect cellular processes like transcription, translation, protein degradation cytoskeleton formation, ion channel properties, and protein signaling. Below are some of the biological processes related to *in vitro* and *in vivo* models of ASD that are involved in ASD risk genes. Some studies have classified the above mentioned proteins and factors on the basis of structural and functional properties that they have (Banerjee, Riordan, and Bhat 2014, Bourgeron 2015, Ebrahimi-Fakhari and Sahin 2015, Liu et al. 2017). Due to space constraints, we will present an overview of synaptic plasticity in autism and related mechanisms, highlighting some important candidate genes based on previous research. Table 1 below has a summary of the factors that we are to discuss.

3.1. Transcription Factors and Chromatin Remodelling

Studies have reported that individuals with ASD have mutations in genes such as methyl-CpG-binding protein 2 (MECP2), monocyte specific enhancer factor 2C (MEF2C), histone deacetylase 4 (HDAC4), chromodomain-helicase-DNA-binding protein 8 (CHD8) and catenin β1 (CTNNB1), which are encoding regulators of chromatin remodeling and gene transcription. Researchers have reported that a subset of these genes regulated by neuronal activity affect neuronal connectivity and also synaptic plasticity.

MECP2 is a protein involved in chromatin remodeling. Using systematic gene screening, it was found that a deletion of, or a point mutation in, the X-linked MECP2 gene causes Rett syndrome (Amir et al. 1999). Rett syndrome occurs in females, due to its X-linked dominant mutation, which is lethal in hemizygous conditions (Sbardella et al. 2017). MECP2 is ubiquitous and is essential for normal neuronal functioning and is important for mature neurons in particular (Shahbazian and Zoghbi 2002). MECP2 is involved in repressing or silencing thousands of other genes. Downregulation of MECP2 results in alterations in global chromatin structure and increased levels of histone acetylation. Phosphorylation of MECP2 results in a reduction in long gene transcription (Gabel et al. 2015). When MECP2 is absent these long genes are depressed (Cohen et al. 2011). Human studies have shown that patient suffering from Rett syndrome have abnormal morphology with decreased dendritic processes, less density of dendritic spine and deformed spine morphology.

Table 1. In this table the ASD linked genes and proteins were classified according to their most common function. The references cited here are either the first report or most useful review

Class	Gene	Name/ Protein	Related function	Other	Reference
Transcription factors and chromatin remodeling	*MECP2*	Methyl-CpG-binding protein 2	Histone methylation, chromatin remodelling, transcriptional repression	Rett syndrome	(Hensch 2004)
	MEF2C	monocyte specific enhancer factor 2C	transcriptional regulation	Severe form of Autism	(Keller et al. 2017)
	HDAC4	histone deacetylase 4	Chromatin remodeling	ASD	(Amir et al. 1999, Gabel et al. 2015, Shahbazian and Zoghbi 2002, Yazdani et al. 2012)
	CHD8	chromodomain-helicase-DNA-bindig protein 8	transcription factors	Early brain development	(Barbosa et al. 2008, Tsai et al. 2012)
	CDKL5	Cyclin-dependent kinases like -5	Synaptic plasticity, gene expression in nucleus	Rett syndrome	(Williams et al. 2010)
	CTNNB1	catenin b1	transcription factors	Early brain development	(Bernier et al. 2014)
Protein turnover (synthesis and degradation) and Signaling	*FMR1*	Fragile X mental retardation 1/FMRP	mRNA binding, maturation, transport to cytosol	ASD	(Shahbazian, Sun, and Zoghbi 2002, Tao et al. 2004)
	NF1	neurofibromin 1	Negative regulator of mTOR, increase in translation in neurons and at the synapse pathway	ADHD/ASD	(O'Roak et al. 2012)
	PTEN	Phosphatase and tensin homolog		ASD	(Verkerk et al. 1991)
	TSC1/TSC2	Tuberous Sclerosis 2	Negative regulator of mTOR but Reduced protein synthesis	ASD	(Johannessen et al. 2005)
Synaptic receptors and cell adhesion	*NRXN1-3*	Neurexins	Majourely Pre-synaptic CAMs essential for synaptic maintenance	Strong ASD candidate	(Lyu et al. 2016)

Table 1. (Continued)

Class	Gene	Name/ Protein	Related function	Other	Reference
molecules (CAM)	*CNTNs/ CNTNAP*	Contactins and Contactin associated proteins (Neurexin subfamily)	GPI anchored Ig-CAM essential for synapse formation and plasticity	ASD	(Serajee et al. 2003)
	GRIN2A/ GRIN2B	Glutamte receptor, ionotropic, N-Methyl D-Asparted 2A/2B	Glutamate neurotransmission	ASD	(Marshall et al. 2008, Zahir et al. 2008)
	CHDs	Cadherins	Transmembranne protein	ASD	(Fernandez et al. 2004, Roohi et al. 2009)
	NLGNs	Neurolignins	Post-synaptic CAMs essential for synaptic maintenece	ASD	(Barnby et al. 2005, O'Roak et al. 2012)
Ion Channels	*SCN1A/SCN2A*	Sodium channel alpha subunit	Sensory perception	ASD	(Bernier et al. 2014)
	CACNA1C	Calcium channel alpha 1C subunit	Activity dependent transcription regulation	ASD	(Jamain et al. 2003)
	KCNs	Potasium channel	Circadian rhythm regulation	ASD	(Weiss et al. 2003)
Scaffolding and cytoskeleton proteins	*SHANKs*	SH3 and multiple ankyrin repeat domains 3	Scaffolding in PSD	ASD, Pheln-McDermit syndrome	(Splawski et al. 2004)
	SYNGAP1	synaptic RAS GTPase-activating protein 1	Synaps development	Early brain development, ASD	(Neale et al. 2012)
Cytoskeleton proteins	*KATNAL2*	katanin-like 2 protein isoforms	Microtubele associated protein		(Yi et al. 2016, Zwanenburg et al. 2016)
Other	RELN	reelin protein	Neuronal migration during brain development	ASD	(Jeyabalan and Clement 2016)
	SYN1	Synapsin 1	Synaptogenesis and neurotransmeter release	ASD	(Neale et al. 2012)
	GABRARs	GABA receptors and its subunits	GABA neurotransmission	ASD *de novo*	(Lammert and Howell 2016)

MEF2C is a protein involved in transcriptional regulation that enables learning and memory. Human studies show that mutations of the MEF2C gene are associated with severe forms of autism (Barbosa et al. 2008). The results of this experimental study in mice suggests that MEF2C does not allow excessive formation of new synapses during activity-dependent refinement of synaptic connections and in doing this, facilitates normal hippocampus-dependent learning and memory. Another study reports that MEF2C induces excitatory synapse elimination in a mouse model in a process that is dependent upon fragile X mental retardation protein. Therefore, patients with mutations in MEF2C and those with fragile X syndrome both have similar increases in synaptic density that occurs through a common mechanism (Pfeiffer et al. 2010, Tsai et al. 2012).

Transcription factors that encode ASD-like genes, CHD8 and CTNNB1, are involved during early development. It has been reported that CHD8 expression increases during the early prenatal period (9-16 weeks post conception) and decreases postnatally (Bernier et al. 2014). In mammals, the developing brain has conserved CHD8 target regions that are related with ASD risk genes. CHD8 acts as a negative regulator of the WNT signaling pathway by suppressing CTNNB1 activity. Both CHD8 and CTNNB1 mutations have different effects on patients; CHD8 patients have larger heads compared to normal individuals, whereas CTNNB1 patients have microcephaly. Studies in zebrafish show that downregulation of CHD8 expression results in the formation of larger brain. Transgenic mice studies show that upregulation of CTNNB1 expression in neural precursors results in increased cerebral cortical surface area and folding. This strongly suggests that, in scenarios where CHD8 and/or CTNNB1 affect brain size, it is because of changes that occurred during proliferation of neuronal progenitors in the brain (Bernier et al. 2014, Chenn and Walsh 2002, O'Roak et al. 2012).

3.2. Protein Turnover (Synthesis and Degradation)

Studies have shown that protein synthesis and degradation are essential for synapse remodeling. Therefore, any change in synaptic proteins influences overall neuronal activity. Kang and Schuman (1996) discovered that local protein synthesis in rat hippocampal slices is required for synaptic plasticity induced by brain-derived neurotrophic factor (BDNF), and in contrast to other types of plasticity, they proposed that this BDNF-induced plasticity requires protein synthesis. Protein synthesis can happen via several pathways but the extracellular signal-regulated kinase (ERK)–mTOR pathway and the fragile X mental retardation protein (FMRP)–eukaryotic translation initiation 4E (EIF4E)–cytoplasmic FMRP-interacting protein 1 (CYFIP1) pathway are the two major regulators of translation. There are several proteins in these pathways that are crucial for brain development and there are several mutation in these genes which result in ASD (Kelleher and Bear 2008). In fact, BDNF and glutamate increase the translation of mRNA by inducing a series of phosphorylation on the mammalian target of rapamycine (mTOR) and fragile X mental retardation protein 1 (FMRP1) pathway. This newly translated protein supports the increase in local synaptic plasticity and the number of local synapses.

There are many genes that play an important part in such activity-driven regulation of synaptic proteins and, in ASD cases, these tend to be mutated (Kelleher and Bear 2008). Global mRNA translation is controlled by the mTOR pathway, and any changes in this pathway results in diseases that are associated with upregulation of cell proliferation and loss of autophagy, including cancer (Kelleher and Bear 2008). These changes also increase the risk of ASD. Genes involved in encoding mTOR pathway repressors cause an increase in neuronal translation, particularly at synapses (e.g., neurofibromin (NF1), PTEN and synaptic RAS GTPase-activating protein 1 (SYNGAP1). Interestingly, another repressor of mTOR, tuberous sclerosis 2 (*TSC2*), if mutated leads to reduced protein synthesis in the hippocampus. Mutations that affect the FMRP–eukaryotic translation initiation 4E (EIF4E)–cytoplasmic FMRP-interacting protein 1

(CYFIP1) complex cause fragile X syndrome and increase the risk of ASD. Local translation of mRNA at the synapse is controlled by this protein complex and this occurs downstream of the RAS-ERK signaling pathway. These mutations were shown to increase mRNA translation (Santini et al. 2013). Inhibition of the signaling factors RAS or ERK activity fixes the unnecessary protein synthesis observed in the fragile X mouse model, whereas the compound rapamycin, a mTOR inhibitor, does not. The translation of more than 1,000 genes is regulated by the FMRP–EIF4E–CYFIP1 complex, several of which are ASD-risk genes. Thus, any changes in the expression levels of the FMRP–EIF4E– CYFIP1 complex results in an imbalance in the levels of essential synaptic proteins that are linked with ASD (Santini et al. 2013).

For protein degradation and recycling, neurons rely heavily on the ubiquitin–proteasome system (UPS), which subsequently affects the regulation of synapse composition, assembly and elimination (Ma and Blenis 2009). Angelman syndrome (a genetic disorder with severe intellectual disability) is a disease in which the gene encoding ubiquitin–protein ligase E3A (UBE3A) is duplicated, resulting in loss of Ube3a-dependent ubiquitination and proteasomal degradation (Greer et al. 2010). In normal conditions, activity-regulated cytoskeleton-associated protein (ARC) decreases long-term potentiation (LTP) by promoting the internalization of AMPA receptors (AMPARs). Disruption of UBE3A function in neurons leads to an increase in ARC expression and an unavoidable decrease in the number of AMPARs at excitatory synapses, turning them to silent synapses.

In summary, whilst not a rule, a lot of mutations have been identified in mTOR, the FMRP–EIF4E–CYFIP and the UPS pathways that lead to a phenotype with an abnormally elevated level of synaptic proteins and an increase in synaptic density. This has been reported in several animal models and in patients with ASD (Ma and Blenis 2009, Santini et al. 2013, Tang et al. 2014).

3.3. Synaptic Receptors and Cell Adhesion Molecules

Many genes and their products that are found to be associated with the increase risk of ASD are involved in neural conduction/transmission. These include receptors such as ionotropic glutamate receptor NMDA 2B (GluN2B), GABAergic receptors (for example, $GABA_A$ receptor subunit α3 (GABRA3 and β3) and glycinergic receptors (for example, glycine receptor subunit α2). Ionic permeability across cell membranes is dependent on proteins, for example, the voltage-dependent calcium channel subunit α1 (CACNA1) protein, the CACNA2 δ3 (CACNA2D3) protein and the sodium channel protein type 1 subunit α (SCN1A) protein. Neurexins (NRXNs) and neurolignins (NLGNs) are synaptic cell adhesion molecules (Sudhof 2008) that are directly involved in the activity-dependent plasticity. Genes like contactin 6 (CNTN6) are involved in neurotransmission and neuritogenesis.

NRXNs and NLGNs are proteins that are located on the presynapse and postsynapse, respectively; they are a binding couple (Sudhof 2008). These genes undergo widespread alternative splicing that can potentially produce up to 3,000 putative isoforms with explicit biochemical and synaptic functions (Sudhof 2008). Calcium/calmodulin-dependent kinase IV (CaMKIV) signalling is known to modulate alternative splicing of NRXN, which is driven by neuronal activity (Sudhof 2008). NLGN subtypes are found at the postsynaptic membrane of glutamatergic, GABAergic and glycinergic synapses. Interestingly, the intracellular carboxy-terminal part of the NLGNs binds to several scaffolding proteins of the postsynaptic density that are encoded by certain ASD-risk genes (Iijima et al. 2011).

Indeed, inhibition of NMDA receptors or the downstream signaling molecule CaMKII activates NLGN1, which results in the suppression of the formation of glutamatergic synapses. Conversely, inhibition of synaptic activity suppresses the formation of GABAergic synapses through the activity of NLGN2 (Chubykin et al. 2007). NLGNs and NRXNs may regulate local synaptic balance (Kwon et al. 2012). Both the NLGN3 knockout and the fragile x syndrome mice models show the extensive synapse formation and decreased metabotropic glutamate receptor-dependent synaptic plasticity (Baudouin et al. 2012).

3.4. Scaffolding Proteins and the Actin Cytoskeleton

The arrangement of neurotransmitter receptors and cell-adhesion molecules at synapses have an intricate architecture of postsynaptic proteins, comprising of signaling molecules, scaffolding proteins and cytoskeletal proteins (Choquet and Triller 2013, Sheng and Kim 2011). Most of these proteins are associated with ASD risk genes. Individuals with ASD were frequently reported to have deletions, duplications and coding mutations in the three SH3 and multiple ankyrin repeat domains protein (SHANK) genes (SHANK1, SHANK2 and SHANK3) (Leblond et al. 2014). SHANKs are thought to interact with glutamate receptors and actin associated proteins by assembling into large molecular podiums. An *in-vitro* study has revealed that SHANK 3 mutations (identified in autism) affect the development and morphology of dendrites as well as axonal growth cone motility via actin accumulation in spines (Durand et al. 2012). Individuals with the 22q13 deletion (including SHANK3 deletion) showed key abnormalities in excitatory but not inhibitory synaptic transmission of neurons derived from their induced pluripotent stem cell. A study on SHANK3 deleted mice reported decreased spine density in the hippocampus concomitant with increased dendritic arborizations in striatal neurons (Shcheglovitov et al. 2013). Studies have found a behaviour resemblance in mice with deleted SHANK genes and humans affected with ASD (Bozdagi et al. 2010, Peca et al. 2011, Wang, McCoy, et al. 2011, Yang et al. 2011). Mice with a SHANK 3 mutation display increased anxiety-like behaviour, decreased vocal communication, decreased locomotion and, extremely enhanced working memory yet decreased long-term memory. In addition, these recent studies on SHANK 3- knockout mice also showed self-harming repetitive grooming along with deficits in social behaviour.

3.5. Synaptic Homeostasis and Excitation/Inhibition (E/I) Balance in ASD

Multiple examples involving rare or *de novo* mutations that have been recognized in ASD patients have been presented above. Mice with mutations in such genes exhibit behavior resembling that of ASD patients, for example, deficits in social behavior and self-harmful repetitive grooming (Bozdagi et al. 2010, Peca et al. 2011, Wang, McCoy, et al. 2011, Yang et al. 2011). (see (Sudhof 2008) for a review). However, apparent behavioral changes due to gene mutation is not always similar between animals and humans. Therefore, the strong phenotype observed in rodents could just be important when discerning the biological roles of ASD-risk associated genes. The behavioral alterations observed in these models, even in adult animals, have been ameliorated by pharmacological intervention (see (Ting, Peca, and Feng 2012) for a review). Improvement of behavioral phenotype in mice with ASD-risk gene mutations with a late onset drug treatment is, in fact, a very hopeful sign for future treatment of humans. Nevertheless, it has yet to be explored whether these drugs could lead to positive results in human ASD patients.

There are two functions that the properties of ASD-risk genes and the different ASD-associated pathways share. Firstly, they modulate synaptic strength and number by partaking in synaptic plasticity. Secondly, in transgenic animal models, these genes affect neuronal networks by either increasing or decreasing synaptic activity or density. Increased neuronal activity results in increased gene transcription and mRNA translation, which is concomitant with the mutations in a majority of ASD related genes, particularly MEF2C, FMR1, NF1, PTEN, SYNGAP1, EIF4E and CYFIP1. These mutations then result in an abnormal augmentation in the synaptic strengthening and number in particular neuronal connections. This abnormal synaptic strength and number may relate to the ectopic synapses observed in the mouse models of ASD (Baudouin et al. 2012) and brains of ASD patients including those with higher risk of epilepsy (SCN1A and SCN2A mutations) (Ey, Leblond, and Bourgeron 2011). Numerous mechanisms are required for the homeostatic regulation of synapses to avert the discrepancy in neuronal connections. These include synaptic scaling (Turrigiano et al. 1998) and metaplasticity (Delorme et al. 2013). In addition to synaptic regulation, changes in neuronal excitability is another mechanism for regulating network activity (Delorme et al. 2013). Synaptic scaling, demonstrated for the first time in central synapses, is when minute changes in the accretion of postsynaptic glutamate receptors occurs with no change in presynaptic function. This is due to increased excitatory synaptic transmission resulting from blockade of activity (Turrigiano et al. 1998). However, studies agree on the fact that a change in AMPAR accumulation leads to synaptic scaling, though in terms of newly accumulated receptors there is disparity in the compositions of their subunits. The voltage gated calcium channels and proteins, such as cytokine tumor necrosis factor, BDNF and ARC, form the signaling pathways for synaptic scaling process (Turrigiano et al. 1998). Metaplasticity is another synaptic regulatory process and is likely due to variations in the ability to induce subsequent synaptic plasticity, such as LTP or long-term depression (LTD). For instance, reduced cortical or hippocampal activity for a certain time period enhances LTP with reticent LTD (Abraham 2008, Abraham and Bear 1996). The molecular basis for this metaplasticity could be upregulation in GluN2B, an NMDAR subunit encoded by ASD-risk associated gene GRIN2B (Chen and Bear 2007, Yang et al. 2011). A second perspective is the downregulation of GABergic inhibition is also regulated by the ASD-risk genes GABRA3 and GABRB3 (Carlson, Wang, and Alger 2002, Chevaleyre and Castillo 2004).

The homeostasis maintained by the above mentioned E/I balance is essential to prevent a possible epileptogenic state, which may be induced if the neuronal population goes out of E/I balance. A recent study suggests that an important condition for synaptic homeostasis is sleep (Tononi and Cirelli 2003). Surprisingly, in this sleep-dependent synaptic homeostasis, NLGNs have a role to play in both flies and rodents. They regulate the levels of glutamatergic and GABAergic currents after sleep deprivation (El Helou et al. 2013, Gilestro, Tononi, and Cirelli 2009, Huber et al. 2004, Li et al. 2013). This might lead to the abnormal sleep and circadian rhythms evident in a majority of ASD patients (Bourgeron 2007).

BDNF release from postsynaptic neurons was shown to be essential for homeostatic retrograde enhancement of presynaptic function (Mozzachiodi and Byrne 2010). Some studies (Halepoto, Bashir, and L 2014, Kasarpalkar, Kothari, and Dave 2014), have reported that an increase in enhancement of presynaptic function through retrospective homeostasis is mediated by BDNF release from postsynaptic neurons (Jakawich et al. 2010). Recently, elevated BDNF levels were found in the blood of ASD patients (Halepoto, Bashir, and L 2014, Kasarpalkar, Kothari, and Dave 2014). Additionally, high plasma levels of serotonin and N-acetylserotonin and low plasma levels of melatonin was frequently reported in ASD patients but not controls (Melke et al. 2008, Pagan et al. 2014). N-acetylserotonin is an agonist for TRKB (tropomysin-related kinaseB, NTRK2), a BDNF receptor; thus, an increase of N-acetylserotonin in the plasma leads to higher protein synthesis due to enhancement of TRKB-induced PI3K (phosphoinositide 3-kinase) signaling (Jang et al. 2010). This effect on protein synthesis resembles the mutation-affected signaling regulation of the mTOR pathway.

4. Conclusion

To summarize, the aberration in neuronal activity in a subset of ASD patients with genetic mutations might be due to a failure of synaptic homeostasis to counteract the effects of these mutations on synaptic activity. Further, mutation-induced abnormal synaptic pruning could be the basis of abnormal synchronization and competition among neuronal connectivity. This could be a possible explanation for problems observed in individuals with ASD in integrating information across different sensory modalities (Rodrigues et al. 2014). This anomaly among sensory stimuli is apparent particularly in the crucial phase of cognitive development during which a firm equilibrium between sensory signals is vital for normal development (Jakawich et al. 2010). In another aspect, a shared presence of autism with epilepsy is also important and must be addressed, however briefly. It has been observed that ASD and epilepsy co-occur frequently in many genetic disorders, such as tuberous sclerosis, Rett syndrome and Fragile X Syndrome. It has been hypothesized that the role of genetic factors in ASD development may also influence the development of epilepsy. This hypothesis gained support with the discovery that mutations of several genes important for neuronal migration and synaptogenesis, such as RELN and SYN1, were found in children with epilepsy, ASD, or both (Marco et al. 2011). Keller and colleagues have covered this topic in detail in a very interesting article (Keller et al. 2017). In the same line, we know that for proper brain function, the balance between inhibition and excitation during neural circuit establishment and employment is absolutely essential..

An imbalance in E/I ratio is often the cause of gene mutations that facilitate synaptogenesis and/or synaptic maintenance via structural scaffolding, transcriptional and translational control and cell adhesion and signalling. Non-compensatory disruption of synaptic E/I equilibrium is often seen in cases of ASD, epilepsy, Rett syndrome, and fragile x syndrome. The brain development also get affected if neural circuit is impaired. The failure of compensatory mechanisms and/or disease-linked homeostatic deficiency impair neural circuits by influencing connectivity and transmissive circuit properties.From a genetic perspective, the individualistic approach is also gaining momentum as the diversity of ASD phenotypes (the spectrum) may be linked to the many different mutations linked to these disorders but also to the ability of each individual's genetic background to buffer the effect of their mutations on synaptic function and each individual's capacity to maintain synaptic homeostasis. This human diversity should be considered the main factor contributing to ASD and autism should be considered not as a single entity, but as a continuum of human heterogeneity.

ACKNOWLEDGMENTS

Authors are thankful to Dr. Arpita Sharma and Vijay Kumar for their valuable feedback on this manuscript.

REFERENCES

Abraham, W. C. (2008). "Metaplasticity: tuning synapses and networks for plasticity." *Nat Rev Neurosci*, *9* (5), 387. doi: 10.1038/nrn2356.

Abraham, W. C. & Bear, M. F. (1996). "Metaplasticity: the plasticity of synaptic plasticity." *Trends Neurosci*, *19* (4), 126-30.

Amir, R. E., Van den Veyver, I. B., Wan, M., Tran, C. Q., Francke, U. & Zoghbi, H. Y. (1999). "Rett syndrome is caused by mutations in X-linked MECP2, encoding methyl-CpG-binding protein 2." *Nat Genet*, *23* (2), 185-8. doi: 10.1038/13810.

An, J. Y. & Claudianos, C. (2016). "Genetic heterogeneity in autism: From single gene to a pathway perspective." *Neurosci Biobehav Rev*, *68*, 442-453. doi: 10.1016/j.neubiorev.2016.06.013.

Anney, R., Klei, L., Pinto, D., Almeida, J., Bacchelli, E., Baird, G., Bolshakova, N., Bolte, S., Bolton, P. F., Bourgeron, T., Brennan, S., Brian, J., Casey, J., Conroy, J., Correia, C., Corsello, C., Crawford, E. L., de Jonge, M., Delorme, R., Duketis, E., Duque, F., Estes, A., Farrar, P., Fernandez, B. A., Folstein, S. E., Fombonne, E., Gilbert, J., Gillberg, C., Glessner, J. T., Green, A., Green, J., Guter, S. J., Heron, E. A., Holt, R., Howe, J. L., Hughes, G., Hus, V., Igliozzi, R., Jacob, S., Kenny, G. P., Kim, C., Kolevzon, A., Kustanovich, V., Lajonchere, C. M., Lamb, J. A., Law-Smith, M., Leboyer, M., Le Couteur, A., Leventhal, B. L., Liu, X. Q., Lombard, F., Lord, C., Lotspeich, L., Lund, S. C., Magalhaes, T. R., Mantoulan, C., McDougle, C. J., Melhem, N. M., Merikangas, A., Minshew, N. J., Mirza, G. K., Munson, J., Noakes, C., Nygren, G., Papanikolaou, K., Pagnamenta, A. T., Parrini, B., Paton, T., Pickles, A., Posey, D. J., Poustka, F., Ragoussis, J., Regan, R., Roberts, W., Roeder, K., Roge, B., Rutter, M. L., Schlitt, S., Shah, N., Sheffield, V. C., Soorya, L., Sousa, I., Stoppioni, V., Sykes, N., Tancredi, R., Thompson, A. P., Thomson, S., Tryfon, A., Tsiantis, J., Van Engeland, H., Vincent, J. B., Volkmar, F., Vorstman, J. A., Wallace, S., Wing, K., Wittemeyer, K., Wood, S., Zurawiecki, D., Zwaigenbaum, L., Bailey, A. J., Battaglia, A., Cantor, R. M., Coon, H., Cuccaro, M. L., Dawson, G., Ennis, S., Freitag, C.

M., Geschwind, D. H., Haines, J. L., Klauck, S. M., McMahon, W. M., Maestrini, E., Miller, J., Monaco, A. P., Nelson, S. F., Nurnberger, J. I., Jr. Oliveira, G., Parr, J. R., Pericak-Vance, M. A., Piven, J., Schellenberg, G. D., Scherer, S. W., Vicente, A. M., Wassink, T. H., Wijsman, E. M., Betancur, C., Buxbaum, J. D., Cook, E. H., Gallagher, L., Gill, M., Hallmayer, J., Paterson, A. D., Sutcliffe, J. S., Szatmari, P., Vieland, V. J., Hakonarson, H. and Devlin, B. (2012). "Individual common variants exert weak effects on the risk for autism spectrum disorders." *Hum Mol Genet*, *21* (21), 4781-92. doi: 10.1093/hmg/dds301.

APA. (2013). "Diagnostic and Statis- tical Manual of Mental Disorders (DSM-5)." *American Psychiatric Association*, *5.*

Bailey, C. H., Kandel, E. R. & Harris, K. M. (2015). "Structural Components of Synaptic Plasticity and Memory Consolidation." *Cold Spring Harb Perspect Biol*, *7* (7), a021758. doi: 10.1101/cshperspect.a021758.

Banerjee, S., Riordan, M. & Bhat, M. A. (2014). "Genetic aspects of autism spectrum disorders: insights from animal models." *Front Cell Neurosci*, *8*, 58. doi: 10.3389/fncel.2014.00058.

Barbosa, A. C., Kim, M. S., Ertunc, M., Adachi, M., Nelson, E. D., McAnally, J., Richardson, J. A., Kavalali, E. T., Monteggia, L. M., Bassel-Duby, R. & Olson, E. N. (2008). "MEF2C, a transcription factor that facilitates learning and memory by negative regulation of synapse numbers and function." *Proc Natl Acad Sci U S A*, *105* (27), 9391-6. doi: 10.1073/pnas.0802679105.

Barnby, G., Abbott, A., Sykes, N., Morris, A., Weeks, D. E., Mott, R., Lamb, J., Bailey, A. J., Monaco, A. P. (2005). Consortium International Molecular Genetics Study of Autism. "Candidate-gene screening and association analysis at the autism-susceptibility locus on chromosome 16p: evidence of association at GRIN2A and ABAT." *Am J Hum Genet*, *76* (6), 950-66. doi: 10.1086/430454.

Baudouin, S. J., Gaudias, J., Gerharz, S., Hatstatt, L., Zhou, K., Punnakkal, P., Tanaka, K. F., Spooren, W., Hen, R., De Zeeuw, C. I., Vogt, K. & Scheiffele, P. (2012). "Shared synaptic pathophysiology in syndromic and nonsyndromic rodent models of autism." *Science*, *338* (6103), 128-32. doi: 10.1126/science.1224159.

Bernier, R., Golzio, C., Xiong, B., Stessman, H. A., Coe, B. P., Penn, O., Witherspoon, K., Gerdts, J., Baker, C., Vulto-van Silfhout, A. T., Schuurs-Hoeijmakers, J. H., Fichera, M., Bosco, P., Buono, S., Alberti, A., Failla, P., Peeters, H., Steyaert, J., Lelm Vissers, L., Francescatto, H. C., Mefford, J. A., Rosenfeld, T., Bakken, B. J., O'Roak, M., Pawlus, R., Moon, J., Shendure, D. G., Amaral, E. Lein., Rankin, J., Romano, C., de Vries, B. B. A., Katsanis, N. and Eichler, E. E. (2014). "Disruptive CHD8 mutations define a subtype of autism early in development." *Cell*, *158* (2), 263-276. doi: 10.1016/j.cell.2014.06.017.

Betancur, C. (2011). "Etiological heterogeneity in autism spectrum disorders: more than 100 genetic and genomic disorders and still counting." *Brain Res*, *1380*, 42-77. doi: 10.1016/j.brainres.2010.11.078.

Blumberg, Stephen J., Bramlett, Matthew D., Kogan, Michael D., Laura A. Schieve., Jones, Jessica R. & Michael, C. Lu. (2013). "Changes in Prevalence of Parent-reported Autism Spectrum Disorder in School-aged U.S. Children: 2007 to 2011–2012." *National Health Statistics Report*, *65*, 1-11.

Bourgeron, T. (2007). "The possible interplay of synaptic and clock genes in autism spectrum disorders." *Cold Spring Harb Symp Quant Biol*, *72*, 645-54. doi: 10.1101/sqb.2007.72.020.

Bourgeron, T. (2015). "From the genetic architecture to synaptic plasticity in autism spectrum disorder." *Nat Rev Neurosci*, *16* (9), 551-63. doi: 10.1038/nrn3992.

Bozdagi, O., Sakurai, T., Papapetrou, D., Wang, X., Dickstein, D. L., Takahashi, N., Kajiwara, Y., Yang, M., Katz, A. M., Scattoni, M. L., Harris, M. J., Saxena, R., Silverman, J. L., Crawley, J. N., Zhou, Q., Hof, P. R. & Buxbaum, J. D. (2010). "Haploinsufficiency of the autism-associated Shank3 gene leads to deficits in synaptic function, social interaction, and social communication." *Mol Autism*, *1* (1), 15. doi: 10.1186/2040-2392-1-15.

Carlson, G., Wang, Y. & Alger, B. E. (2002). "Endocannabinoids facilitate the induction of LTP in the hippocampus." *Nat Neurosci*, *5* (8), 723-4. doi: 10.1038/nn879.

Chen, W. S. & Bear, M. F. (2007). "Activity-dependent regulation of NR2B translation contributes to metaplasticity in mouse visual cortex." *Neuropharmacology*, *52* (1), 200-14. doi: 10.1016/j.neuropharm.2006.07.003.

Chenn, A. & Walsh, C. A. (2002). "Regulation of cerebral cortical size by control of cell cycle exit in neural precursors." *Science*, *297* (5580), 365-9. doi: 10.1126/science.1074192.

Chevaleyre, V. & Castillo, P. E. (2004). "Endocannabinoid-mediated metaplasticity in the hippocampus." *Neuron*, 43 (6), 871-81. doi: 10.1016/j.neuron.2004.08.036.

Choquet, D. & Triller, A. (2013). "The dynamic synapse." *Neuron*, *80* (3), 691-703. doi: 10.1016/j.neuron.2013.10.013.

Chubykin, A. A., Atasoy, D., Etherton, M. R., Brose, N., Kavalali, E. T., Gibson, J. R. & Sudhof, T. C. (2007). "Activity-dependent validation of excitatory versus inhibitory synapses by neuroligin-1 versus neuroligin-2." *Neuron*, *54* (6), 919-31. doi: 10.1016/j.neuron.2007.05.029.

Cohen, S., Gabel, H. W., Hemberg, M., Hutchinson, A. N., Sadacca, L. A., Ebert, D. H., Harmin, D. A., Greenberg, R. S., Verdine, V. K., Zhou, Z., Wetsel, W. C., West, A. E. & Greenberg, M. E. (2011). "Genome-wide activity-dependent MeCP2 phosphorylation regulates nervous system development and function." *Neuron*, *72* (1), 72-85. doi: 10.1016/ j.neuron.2011.08.022.

Cross-Disorder Group of the Psychiatric Genomics, Consortium. (2013). "Identification of risk loci with shared effects on five major psychiatric disorders: a genome-wide analysis." *Lancet*, *381* (9875), 1371-9. doi: 10.1016/S0140-6736(12)62129-1.

Deelen, W., Bakker, C., Halley, D. J. & Oostra, B. A. (1994). "Conservation of CGG region in FMR1 gene in mammals." *Am J Med Genet*, *51* (4), 513-6. doi: 10.1002/ajmg.1320510445.

Delorme, R., Ey, E., Toro, R., Leboyer, M., Gillberg, C., & Bourgeron, T. (2013). "Progress toward treatments for synaptic defects in autism." *Nat Med*, *19* (6), 685-94. doi: 10.1038/nm.3193.

Dolen, G. & Bear, M. F. (2008). "Role for metabotropic glutamate receptor 5 (mGluR5) in the pathogenesis of fragile X syndrome." *J Physiol*, *586* (6), 1503-8. doi: 10.1113/jphysiol.2008.150722.

Durand, C. M., Perroy, J., Loll, F., Perrais, D., Fagni, L., Bourgeron, T., Montcouquiol, M. & Sans, N. (2012). "SHANK3 mutations identified in autism lead to modification of dendritic spine morphology via an actin-dependent mechanism." *Mol Psychiatry*, *17* (1), 71-84. doi: 10.1038/ mp.2011.57.

Ebrahimi-Fakhari, D. & Sahin, M. (2015). "Autism and the synapse: emerging mechanisms and mechanism-based therapies." *Curr Opin Neurol*, *28* (2), 91-102. doi: 10.1097/WCO.0000000000000186.

El Helou, J., Belanger-Nelson, E., Freyburger, M., Dorsaz, S., Curie, T., La Spada, F., Gaudreault, P. O., Beaumont, E., Pouliot, P., Lesage, F., Frank, M. G., Franken, P. & Mongrain, V. (2013). "Neuroligin-1 links neuronal activity to sleep-wake regulation." *Proc Natl Acad Sci U S A*, *110* (24), 9974-9. doi: 10.1073/pnas.1221381110.

Ey, E., Leblond, C. S. & Bourgeron, T. (2011). "Behavioral profiles of mouse models for autism spectrum disorders." *Autism Res*, *4* (1), 5-16. doi: 10.1002/aur.175.

Fernandez, T., Morgan, T., Davis, N., Klin, A., Morris, A., Farhi, A., Lifton, R. P. & State, M. W. (2004). "Disruption of contactin 4 (CNTN4) results in developmental delay and other features of 3p deletion syndrome." *Am J Hum Genet*, *74* (6), 1286-93. doi: 10.1086/421474.

Gabel, H. W.,. Kinde, B., Stroud, H., Gilbert, C. S., Harmin, D. A., Kastan, N. R., Hemberg, M., Ebert, D. H. & Greenberg, M. E. (2015). "Disruption of DNA-methylation-dependent long gene repression in Rett syndrome." *Nature*, *522* (7554), 89-93. doi: 10.1038/nature 14319.

Gilestro, G. F., Tononi, G. & Cirelli, C. (2009). "Widespread changes in synaptic markers as a function of sleep and wakefulness in

Drosophila." *Science*, *324* (5923), 109-12. doi: 10.1126/ science. 1166673.

Gillberg, C. (2010). "The ESSENCE in child psychiatry: Early Symptomatic Syndromes Eliciting Neurodevelopmental Clinical Examinations." *Res Dev Disabil*, *31* (6), 1543-51. doi: 10.1016/ j.ridd.2010.06.002.

Greer, P. L., Hanayama, R., Bloodgood, B. L., Mardinly, A. R., Lipton, D. M., Flavell, S. W., Kim, T. K., Griffith, E. C., Waldon, Z., Maehr, R., Ploegh, H. L., Chowdhury, S., Worley, P. F., Steen, J. & Greenberg, M. E. (2010). "The Angelman Syndrome protein Ube3A regulates synapse development by ubiquitinating arc." *Cell*, *140* (5), 704-16. doi: 10.1016/j.cell.2010.01.026.

Hagerman, R. J. & Hagerman, P. J. (2002). "The fragile X premutation: into the phenotypic fold." *Curr Opin Genet Dev*, *12* (3), 278-83. doi: 10.1016/S0959-437X(02)00299-X.

Halepoto, D. M., Bashir, S. & Ayadhi, A. L. (2014). "Possible role of brain-derived neurotrophic factor (BDNF) in autism spectrum disorder: current status." *J Coll Physicians Surg Pak*, *24* (4), 274-8. doi: 04.2014/JCPSP.274278.

Hensch, T. K. (2004). "Critical period regulation." *Annu Rev Neurosci*, *27*, 549-79. doi: 10.1146/annurev.neuro.27.070203.144327.

Huber, R., Ghilardi, M. F., Massimini, M. & Tononi, G. (2004). "Local sleep and learning." *Nature*, *430* (6995), 78-81. doi: 10.1038/ nature02663.

Iijima, T., Wu, K., Witte, H., Hanno-Iijima, Y., Glatter, T., Richard, S. & Scheiffele, P. (2011). "SAM68 regulates neuronal activity-dependent alternative splicing of neurexin-1." *Cell*, *147* (7), 1601-14. doi: 10.1016/j.cell.2011.11.028.

Jakawich, S. K., Nasser, H. B., Strong, M. J., McCartney, A. J., Perez, A. S., Rakesh, N., Carruthers, C. J. & Sutton, M. A. (2010). "Local presynaptic activity gates homeostatic changes in presynaptic function driven by dendritic BDNF synthesis." *Neuron*, *68* (6), 1143-58. doi: 10.1016/j.neuron.2010.11.034.

Jamain, S., Quach, H., Betancur, C., Rastam, M., Colineaux, C., Gillberg, I. C., Soderstrom, H., Giros, B., Leboyer, M., Gillberg, C. & Bourgeron, T. (2003). Study Paris Autism Research International Sibpair. "Mutations of the X-linked genes encoding neuroligins NLGN3 and NLGN4 are associated with autism." *Nat Genet*, *34* (1), 27-9. doi: 10.1038/ng1136.

Jang, S. W., Liu, X., Pradoldej, S., Tosini, G., Chang, Q., Iuvone, P. M. & Ye, K. (2010). "N-acetylserotonin activates TrkB receptor in a circadian rhythm." *Proc Natl Acad Sci U S A*, *107* (8), 3876-81. doi: 10.1073/ pnas.0912531107.

Jeyabalan, N. & Clement, J. P. (2016). "SYNGAP1: Mind the Gap." *Front Cell Neurosci*, *10*, 32. doi: 10.3389/fncel.2016.00032.

Johannessen, C. M., Reczek, E. E., James, M. F., Brems, H., Legius, E. & Cichowski, K. (2005). "The NF1 tumor suppressor critically regulates TSC2 and mTOR." *Proc Natl Acad Sci U S A*, *102* (24), 8573-8. doi: 10.1073/pnas.0503224102.

Kang, H. & Schuman, E. M. (1996). "A requirement for local protein synthesis in neurotrophin-induced hippocampal synaptic plasticity." *Science*, *273* (5280), 1402-6.

Kanner, L. (1943). "Autistic disturbances of affective contact." *Nervous Child*, *2*, 217-250.

Kasarpalkar, N. J., Kothari, S. T. & Dave, U. P. (2014). "Brain-Derived Neurotrophic Factor in children with Autism Spectrum Disorder." *Ann Neurosci*, *21* (4), 129-33. doi: 10.5214/ans.0972.7531.210403.

Kelleher, R. J. 3rd. & Bear, M. F. (2008). "The autistic neuron: troubled translation?" *Cell*, *135* (3), 401-6. doi: 10.1016/j.cell.2008.10.017.

Keller, R., Basta, R., Salerno, L. & Elia, M. (2017). "Autism, epilepsy, and synaptopathies: a not rare association." *Neurol Sci*, *38* (8), 1353-1361. doi: 10.1007/s10072-017-2974-x.

Kwon, H. B., Kozorovitskiy, Y., Oh, W. J., Peixoto, R. T., Akhtar, N., Saulnier, J. L., Gu, C. & Sabatini, B. L. (2012). "Neuroligin-1-dependent competition regulates cortical synaptogenesis and synapse number." *Nat Neurosci*, *15* (12), 1667-74. doi: 10.1038/nn.3256.

Lammert, D. B. & Howell, B. W. (2016). "RELN Mutations in Autism Spectrum Disorder." *Front Cell Neurosci*, *10*, 84. doi: 10.3389/ fncel. 2016.00084.

Leblond, C. S., Nava, C., Polge, A., Gauthier, J., Huguet, G., Lumbroso, S., Giuliano, F., Stordeur, C., Depienne, C., Mouzat, K., Pinto, D., Howe, J., Lemiere, N., Durand, C. M., Guibert, J., Ey, E., Toro, R., Peyre, H., Mathieu, A., Amsellem, F., Rastam, M., Gillberg, I. C., Rappold, G. A., Holt, R., Monaco, A. P., Maestrini, E., Galan, P., Heron, D., Jacquette, A., Afenjar, A., Rastetter, A., Brice, A., Devillard, F., Assouline, B., Laffargue, F., Lespinasse, J., Chiesa, J., Rivier, F., Bonneau, D., Regnault, B., Zelenika, D., Delepine, M., Lathrop, M., Sanlaville, D., Schluth-Bolard, C., Edery, P., Perrin, L., Tabet, A. C., Schmeisser, M. J., Boeckers, T. M., Coleman, M., Sato, D., Szatmari, P., Scherer, S. W., Rouleau, G. A., Betancur, C., Leboyer, M., Gillberg, C., Delorme, R. & Bourgeron, T. (2014). "Meta-analysis of SHANK Mutations in Autism Spectrum Disorders: a gradient of severity in cognitive impairments." *PLoS Genet*, *10* (9), e1004580. doi: 10.1371/journal.pgen.1004580.

Li, Y., Zhou, Z., Zhang, X., Tong, H., Li, P., Zhang, Z. C., Jia, Z., Xie, W. & Han, J. (2013). "Drosophila neuroligin 4 regulates sleep through modulating GABA transmission." *J Neurosci*, *33* (39), 15545-54. doi: 10.1523/JNEUROSCI.0819-13.2013.

Liu, X., Li, Z., Fan, C., Zhang, D. & Chen, J. (2017). "Genetics implicate common mechanisms in autism and schizophrenia: synaptic activity and immunity." *J Med Genet*, *54* (8), 511-520. doi: 10.1136/ jmedgenet-2016-104487.

Lyu, J. W., Yuan, B., Cheng, T. L., Qiu, Z. L. & Zhou, W. H. (2016). "Reciprocal regulation of autism-related genes MeCP2 and PTEN via microRNAs." *Sci Rep*, *6*, 20392. doi: 10.1038/srep20392.

Ma, X. M. & Blenis, J. (2009). "Molecular mechanisms of mTOR-mediated translational control." *Nat Rev Mol Cell Biol*, *10* (5), 307-18. doi: 10.1038/nrm2672.

Marco, E. J., Hinkley, L. B., Hill, S. S. & Nagarajan, S. S. (2011). "Sensory processing in autism: a review of neurophysiologic findings."

Pediatr Res, *69* (5 Pt 2), 48R-54R. doi: 10.1203/ PDR. 0b013e3182130c54.

Marshall, C. R., Noor, A., Vincent, J. B., Lionel, A. C., Feuk, L., Skaug, J., Shago, M., Moessner, R., Pinto, D., Ren, Y., Thiruvahindrapduram, B., Fiebig, A., Schreiber, S., Friedman, J., Ketelaars, C. E., Vos, Y. J., Ficicioglu, C., Kirkpatrick, S., Nicolson, R., Sloman, L., Summers, A., Gibbons, C. A., Teebi, A., Chitayat, D., Weksberg, R., Thompson, A., Vardy, C., Crosbie, V., Luscombe, S., Baatjes, R., Zwaigenbaum, L., Roberts, W., Fernandez, B., Szatmari, P. & Scherer, S. W. (2008). "Structural variation of chromosomes in autism spectrum disorder." *Am J Hum Genet*, *82* (2), 477-88. doi: 10.1016/j.ajhg.2007.12.009.

McNaughton, C. H., Moon, J., Strawderman, M. S., Maclean, K. N., Evans, J. & Strupp, B. J. (2008). "Evidence for social anxiety and impaired social cognition in a mouse model of fragile X syndrome." *Behav Neurosci*, *122* (2), 293-300. doi: 10.1037/0735-7044.122.2.293.

Melke, J., Goubran Botros, H., Chaste, P., Betancur, C., Nygren, G., Anckarsater, H., Rastam, M., Stahlberg, O., Gillberg, I. C., Delorme, R., Chabane, N., Mouren-Simeoni, M. C., Fauchereau, F., Durand, C. M., Chevalier, F., Drouot, X., Collet, C., Launay, J. M., Leboyer, M., Gillberg, C. & Bourgeron, T. (2008). "Abnormal melatonin synthesis in autism spectrum disorders." *Mol Psychiatry*, *13* (1), 90-8. doi: 10.1038/sj.mp.4002016.

Modabbernia, A., Velthorst, E. & Reichenberg, A. (2017). "Environmental risk factors for autism: an evidence-based review of systematic reviews and meta-analyses." *Mol Autism*, *8*, 13. doi: 10.1186/s13229-017-0121-4.

Mozzachiodi, R. & Byrne, J. H. (2010). "More than synaptic plasticity: role of nonsynaptic plasticity in learning and memory." *Trends Neurosci*, *33* (1), 17-26. doi: 10.1016/j.tins.2009.10.001.

Neale, B. M., Kou, Y., Liu, L., Ma'ayan, A., Samocha, K. E., Sabo, A., Lin, C. F., Stevens, C., Wang, L. S., Makarov, V., Polak, P., Yoon, S., Maguire, J., Crawford, E. L., Campbell, N. G., Geller, E. T., Valladares, O., Schafer, C., Liu, H., Zhao, T., Cai, G., Lihm, J., Dannenfelser, R., Jabado, O., Peralta, Z., Nagaswamy, U., Muzny, D.,

Reid, J. G., Newsham, I., Wu, Y., Lewis, L., Han, Y., Voight, B. F., Lim, E., Rossin, E., Kirby, A., Flannick, J., Fromer, M., Shakir, K., Fennell, T., Garimella, K., Banks, E., Poplin, R., Gabriel, S., DePristo, M., Wimbish, J. R., Boone, B. E., Levy, S. E., Betancur, C., Sunyaev, S., Boerwinkle, E., Buxbaum, J. D., Cook, E. H., Jr. Devlin, B., Gibbs, R. A., Roeder, K., Schellenberg, G. D., Sutcliffe, J. S. & Daly, M. J. (2012). "Patterns and rates of exonic de novo mutations in autism spectrum disorders." *Nature*, *485* (7397), 242-5. doi: 10.1038/nature 11011.

O'Roak, B. J., Vives, L., Fu, W., Egertson, J. D., Stanaway, I. B., Phelps, I. G., Carvill, G., Kumar, A., Lee, C., Ankenman, K., Munson, J., Hiatt, J. B., Turner, E. H., Levy, R., O'Day, D. R., Krumm, N., Coe, B. P., Martin, B. K., Borenstein, E., Nickerson, D. A., Mefford, H. C., Doherty, D., Akey, J. M., Bernier, R., Eichler, E. E. & Shendure, J. (2012). "Multiplex targeted sequencing identifies recurrently mutated genes in autism spectrum disorders." *Science*, *338* (6114), 1619-22. doi: 10.1126/science.1227764.

Pagan, C., Delorme, R., Callebert, J., Goubran-Botros, H., Amsellem, F., Drouot, X., Boudebesse, C., Le Dudal, K., Ngo-Nguyen, N., Laouamri, H., Gillberg, C., Leboyer, M., Bourgeron, T. & Launay, J. M. (2014). "The serotonin-N-acetylserotonin-melatonin pathway as a biomarker for autism spectrum disorders." *Transl Psychiatry*, *4*, e479. doi: 10.1038/ tp.2014.120.

Peca, J., Feliciano, C., Ting, J. T., Wang, W., Wells, M. F., Venkatraman, T. N., Lascola, C. D., Fu, Z. & Feng, G. (2011). "Shank3 mutant mice display autistic-like behaviours and striatal dysfunction." *Nature*, *472* (7344), 437-42. doi: 10.1038/nature09965.

Pfeiffer, B. E., Zang, T., Wilkerson, J. R., Taniguchi, M., Maksimova, M. A., Smith, L. N., Cowan, C. W. & Huber, K. M. (2010). "Fragile X mental retardation protein is required for synapse elimination by the activity-dependent transcription factor MEF2." *Neuron*, *66* (2), 191-7. doi: 10.1016/j.neuron.2010.03.017.

Rodrigues, D. H., Rocha, N. P., Sousa, L. F., Barbosa, I. G., Kummer, A. & Teixeira, A. L. (2014). "Circulating levels of neurotrophic factors in autism spectrum disorders." *Neuro Endocrinol Lett*, *35* (5), 380-4.

Ronald, A. & Hoekstra, R. A. (2011). "Autism spectrum disorders and autistic traits: a decade of new twin studies." *Am J Med Genet B Neuropsychiatr Genet*, *156B* (3), 255-74. doi: 10.1002/ajmg.b.31159.

Roohi, J., Montagna, C., Tegay, D. H., Palmer, L. E., DeVincent, C., Pomeroy, J. C., Christian, S. L., Nowak, N. & Hatchwell, E. (2009). "Disruption of contactin 4 in three subjects with autism spectrum disorder." *J Med Genet*, *46* (3), 176-82. doi: 10.1136/jmg.2008.057505.

Sandin, S., Lichtenstein, P., Kuja-Halkola, R., Larsson, H., Hultman, C. M. & Reichenberg, A. (2014). "The familial risk of autism." *JAMA*, *311* (17), 1770-7. doi: 10.1001/jama.2014.4144.

Santini, E., Huynh, T. N., MacAskill, A. F., Carter, A. G., Pierre, P., Ruggero, D., Kaphzan, H. & Klann, E. (2013). "Exaggerated translation causes synaptic and behavioural aberrations associated with autism." *Nature*, *493* (7432), 411-5. doi: 10.1038/nature11782.

Sbardella, D., Tundo, G. R., Campagnolo, L., Valacchi, G., Orlandi, A., Curatolo, P., Borsellino, G., D'Esposito, M., Ciaccio, C., Cesare, S. D., Pierro, D. D., Galasso, C., Santarone, M. E., Hayek, J., Coletta, M. & Marini, S. (2017). "Retention of Mitochondria in Mature Human Red Blood Cells as the Result of Autophagy Impairment in Rett Syndrome." *Sci Rep*, *7* (1), 12297. doi: 10.1038/s41598-017-12069-0.

Serajee, F. J., Nabi, R., Zhong, H. & Mahbubul Huq, A. H. (2003). "Association of INPP1, PIK3CG, and TSC2 gene variants with autistic disorder: implications for phosphatidylinositol signalling in autism." *J Med Genet*, *40* (11), 119.

Shahbazian, M. D., Sun, Y. & Zoghbi, H. Y. (2002). "Balanced X chromosome inactivation patterns in the Rett syndrome brain." *Am J Med Genet*, *111* (2), 164-8. doi: 10.1002/ajmg.10557.

Shahbazian, M. D. & Zoghbi, H. Y. (2002). "Rett syndrome and MeCP2: linking epigenetics and neuronal function." *Am J Hum Genet*, *71* (6), 1259-72. doi: 10.1086/345360.

Shcheglovitov, A., Shcheglovitova, O., Yazawa, M., Portmann, T., Shu, R., Sebastiano, V., Krawisz, A., Froehlich, W., Bernstein, J. A., Hallmayer, J. F. & Dolmetsch, R. E. (2013). "SHANK3 and IGF1 restore synaptic deficits in neurons from 22q13 deletion syndrome patients." *Nature*, *503* (7475), 267-71. doi: 10.1038/nature12618.

Sheng, M. & Kim, E. (2011). "The postsynaptic organization of synapses." *Cold Spring Harb Perspect Biol*, *3* (12). doi: 10.1101/cshperspect.a005678.

Sherman, S. L., Marsteller, F., Abramowitz, A. J., Scott, E., Leslie, M. & Bregman, J. (2002). "Cognitive and behavioral performance among FMR1 high-repeat allele carriers surveyed from special education classes." *Am J Med Genet*, *114* (4), 458-65. doi: 10.1002/ajmg.10303.

Sidorov, Michael S., Benjamin, Auerbach, D. & Mark, Bear, F. (2013). "Fragile X mental retardation protein and synaptic plasticity." *Molecular brain*, *6* (1), 15. doi: 10.1186/1756-6606-6-15.

Spencer, C. M., Alekseyenko, O., Hamilton, S. M., Thomas, A. M., Serysheva, E., Yuva-Paylor, L. A. & Paylor, R. (2011). "Modifying behavioral phenotypes in Fmr1KO mice: genetic background differences reveal autistic-like responses." *Autism Res*, *4* (1), 40-56. doi: 10.1002/aur.168.

Splawski, I., Timothy, K. W., Sharpe, L. M., Decher, N., Kumar, P., Bloise, R., Napolitano, C., Schwartz, P. J., Joseph, R. M., Condouris, K., Tager-Flusberg, H., Priori, S. G., Sanguinetti, M. C. & Keating, M. T. (2004). "Ca(V)1.2 calcium channel dysfunction causes a multisystem disorder including arrhythmia and autism." *Cell*, *119* (1), 19-31. doi: 10.1016/j.cell.2004.09.011.

Sudhof, T. C. (2008). "Neuroligins and neurexins link synaptic function to cognitive disease." *Nature*, *455*(7215), 903-11. doi: 10.1038/nature07456.

Suvrathan, A. & Chattarji, S. (2011). "Fragile X syndrome and the amygdala." *Curr Opin Neurobiol*, *21* (3), 509-15. doi: 10.1016/j.conb.2011.04.005.

Suvrathan, A., Hoeffer, C. A., Wong, H., Klann, E. & Chattarji, S. (2010). "Characterization and reversal of synaptic defects in the amygdala in a mouse model of fragile X syndrome." *Proc Natl Acad Sci U S A*, *107* (25), 11591-6. doi: 10.1073/pnas.1002262107.

Tang, G., Gudsnuk, K., Kuo, S. H., Cotrina, M. L., Rosoklija, G., Sosunov, A., Sonders, M. S., Kanter, E., Castagna, C., Yamamoto, A., Yue, Z., Arancio, O., Peterson, B. S., Champagne, F., Dwork, A. J., Goldman, J. & Sulzer, D. (2014). "Loss of mTOR-dependent macroautophagy causes autistic-like synaptic pruning deficits." *Neuron*, *83* (5), 1131-43. doi: 10.1016/j.neuron.2014.07.040.

Tao, J., Van Esch, H., Hagedorn-Greiwe, M., Hoffmann, K., Moser, B., Raynaud, M., Sperner, J., Fryns, J. P., Schwinger, E., Gecz, J., Ropers, H. H. & Kalscheuer, V. M. (2004). "Mutations in the X-linked cyclin-dependent kinase-like 5 (CDKL5/STK9) gene are associated with severe neurodevelopmental retardation." *Am J Hum Genet*, *75* (6), 1149-54. doi: 10.1086/426460.

Ting, J. T., Peca, J. & Feng, G. (2012). "Functional consequences of mutations in postsynaptic scaffolding proteins and relevance to psychiatric disorders." *Annu Rev Neurosci*, *35*, 49-71. doi: 10.1146/annurev-neuro-062111-150442.

Tononi, G. & Cirelli, C. (2003). "Sleep and synaptic homeostasis: a hypothesis." *Brain Res Bull*, *62* (2), 143-50.

Tsai, N. P., Wilkerson, J. R., Guo, W., Maksimova, M. A., DeMartino, G. N., Cowan, C. W. & Huber, K. M. (2012). "Multiple autism-linked genes mediate synapse elimination via proteasomal degradation of a synaptic scaffold PSD-95." *Cell*, *151* (7), 1581-94. doi: 10.1016/j.cell.2012.11.040.

Turner, G., Webb, T., Wake, S. & Robinson, H. (1996). "Prevalence of fragile X syndrome." *Am J Med Genet*, *64* (1), 196-7. doi: 10.1002/(sici)1096-8628(19960712)64:1<196::aid-ajmg35>3.0.co;2-g.

Turrigiano, G. G., Leslie, K. R., Desai, N. S., Rutherford, L. C. & Nelson, S. B. (1998). "Activity-dependent scaling of quantal amplitude in neocortical neurons." *Nature*, *391* (6670), 892-6. doi: 10.1038/36103.

Ung, D. C., Iacono, G., Meziane, H., Blanchard, E., Papon, M. A., Selten, M., van Rhijn, J. R., Montjean, R., Rucci, J., Martin, S., Fleet, A., Birling, M. C., Marouillat, S., Roepman, R., Selloum, M., Lux, A.,

Thepault, R. A., Hamel, P., Mittal, K., Vincent, J. B., Dorseuil, O., Stunnenberg, H. G., Billuart, P., Nadif Kasri, N., Herault, Y. & Laumonnier, F. (2017). "Ptchd1 deficiency induces excitatory synaptic and cognitive dysfunctions in mouse." *Mol Psychiatry*. doi: 10.1038/mp.2017.39.

Verkerk, A. J., Pieretti, M., Sutcliffe, J. S., Fu, Y. H., Kuhl, D. P., Pizzuti, A., Reiner, O., Richards, S., Victoria, M. F., Zhang, F. P., et al. (1991). "Identification of a gene (FMR-1) containing a CGG repeat coincident with a breakpoint cluster region exhibiting length variation in fragile X syndrome." *Cell*, *65* (5), 905-14.

Wang, F., Kessels, H. W. & Hu, H. (2014). "The mouse that roared: neural mechanisms of social hierarchy." *Trends Neurosci*, *37* (11), 674-82. doi: 10.1016/j.tins.2014.07.005.

Wang, F., Zhu, J., Zhu, H., Zhang, Q., Lin, Z. & Hu, H. (2011). "Bidirectional control of social hierarchy by synaptic efficacy in medial prefrontal cortex." *Science*, *334* (6056), 693-7. doi: 10.1126/science.1209951.

Wang, X., McCoy, P. A., Rodriguiz, R. M., Pan, Y., Je, H. S., Roberts, A. C., Kim, C. J., Berrios, J., Colvin, J. S., Bousquet-Moore, D., Lorenzo, I., Wu, G., Weinberg, R. J., Ehlers, M. D., Philpot, B. D., Beaudet, A. L., Wetsel, W. C. & Jiang, Y. H. (2011). "Synaptic dysfunction and abnormal behaviors in mice lacking major isoforms of Shank3." *Hum Mol Genet*, *20* (15), 3093-108. doi: 10.1093/hmg/ddr212.

Weiss, L. A., Escayg, A., Kearney, J. A., Trudeau, M., MacDonald, B. T., Mori, M., Reichert, J., Buxbaum, J. D. & Meisler, M. H. (2003). "Sodium channels SCN1A, SCN2A and SCN3A in familial autism." *Molecular psychiatry*, *8* (2), 186.

Williams, S. R., Aldred, M. A., Der Kaloustian, V. M., Halal, F., Gowans, G., McLeod, D. R., Zondag, S., Toriello, H. V., Magenis, R. E. & Elsea, S. H. (2010). "Haploinsufficiency of HDAC4 causes brachydactyly mental retardation syndrome, with brachydactyly type E, developmental delays, and behavioral problems." *Am J Hum Genet*, *87* (2), 219-28. doi: 10.1016/j.ajhg.2010.07.011.

Yang, Q., Liao, Z. H., Xiao, Y. X., Lin, Q. S., Zhu, Y. S. & Li, S. T. (2011). "Hippocampal synaptic metaplasticity requires the activation of NR2B-containing NMDA receptors." *Brain Res Bull*, *84* (2), 137-43. doi: 10.1016/j.brainresbull.2010.12.009.

Yazdani, M., Deogracias, R., Guy, J., Poot, R. A., Bird, A. & Barde, Y. A. (2012). "Disease modeling using embryonic stem cells: MeCP2 regulates nuclear size and RNA synthesis in neurons." *Stem Cells*, *30* (10), 2128-39. doi: 10.1002/stem.1180.

Yi, F., Danko, T., Botelho, S. C., Patzke, C., Pak, C., Wernig, M. & Sudhof, T. C. (2016). "Autism-associated SHANK3 haploinsufficiency causes Ih channelopathy in human neurons." *Science*, *352* (6286), aaf2669. doi: 10.1126/science.aaf2669.

Zahir, F. R., Baross, A., Delaney, A. D., Eydoux, P., Fernandes, N. D., Pugh, T., Marra, M. A. & Friedman, J. M. (2008). "A patient with vertebral, cognitive and behavioural abnormalities and a de novo deletion of NRXN1alpha." *J Med Genet*, *45* (4), 239-43. doi: 10.1136/jmg.2007.054437.

Zwanenburg, R. J., Ruiter, S. A., van den Heuvel, E. R., Flapper, B. C. & Van Ravenswaaij-Arts, C. M. (2016). "Developmental phenotype in Phelan-McDermid (22q13.3 deletion) syndrome: a systematic and prospective study in 34 children." *J Neurodev Disord*, *8*, 16. doi: 10.1186/s11689-016-9150-0.

In: Synaptic Plasticity
Editor: Sraboni Chaudhury

ISBN: 978-1-53613-606-7

Chapter 6

HIPPOCAMPAL SYNAPTIC PLASTICITY IMPAIRMENT IN FRAGILE-X SYNDROME

Joana Gil-Mohapel, PhD
Island Medical Program, University of British Columbia and Division of Medical Sciences, University of Victoria, Victoria, BC, Canada

ABSTRACT

Fragile-X Syndrome (FXS) is the most common form of inherited intellectual disability and the leading genetic cause of autism spectrum disorders. FXS is caused by transcriptional silencing of the Fragile X Mental Retardation 1 (*Fmr1*) gene due to a cytosine-guanine-guanine (CGG) repeat expansion, resulting in the loss of Fragile X Mental Retardation Protein (FMRP). FMRP is involved in transcriptional regulation and trafficking of mRNA from the nucleus to the cytoplasm and distal sites both in pre- and post-synaptic terminals. Consequently, FXS is a multifaceted disorder associated with impaired synaptic plasticity. One region of the brain that is significantly impacted by the loss of FMRP is the hippocampus, a structure that plays a critical role in the regulation of mood and cognition. This book chapter provides an overview of how synaptic deficits in hippocampal sub-regions, including the *cornu ammonis* (CA) 1 and the dentate gyrus (DG) sub-fields, contribute to cognitive impairments associated with this neurodevelopmental disorder.

Keywords: *Fmr1* knock-out mice, Fragile-X Syndrome, hippocampus, synaptic plasticity, NMDA receptor, mGluR theory

1. Introduction

Fragile-X Syndrome (FXS) affects approximately 1 in 4000 males and 1 in 8000 females making it one of the most common forms of cognitive impairment behavior (Turner et al., 1996). FXS is caused by transcriptional repression of the Fragile X Mental Retardation 1 (*Fmr1*) gene encoding the Fragile X Mental Retardation Protein (FMRP). In FXS patients, cytosine-guanine-guanine (CGG) trinucleotide expansions of over 200 repeats (Fu et al., 1991) cause the hypermethylation and transcriptional silencing of the *Fmr1* gene and loss of FMRP (Pieretti et al., 1991, Sutcliffe et al., 1992, Hornstra et al., 1993, O'Donnell and Warren, 2002). FMRP trafficks mRNA from the nucleus to the cytoplasm and distal sites both in pre- and postsynaptic terminals, regulating the translation of multiple proteins, including several that are important for learning and memory (Ashley et al., 1993, Antar and Bassell, 2003, Christie and Cameron, 2006, Bassell and Warren, 2008).

FXS is associated with a number of changes in the appearance and behavior of individuals. The characteristic physical phenotype of FXS includes an elongated and narrow face, large or protruding ears, and a prominent forehead and jaw (Chudley and Hagerman, 1987, Verma and Elango, 1994). Other features include connective tissue dysplasia such as a high-arched palate (Hagerman et al., 1984, Partington, 1984, Hagerman and Hagerman, 2002), hyper-extendable joints (Hagerman et al., 1984), and soft, velvet-like skin (Turner et al., 1980). The clinical severity of FXS correlates with the degree of *Fmr1* methylation (Reiss et al., 1995, Hagerman and Hagerman, 2002), so often the physical features associated with FXS are typically less pronounced in heterozygous females than homozygous males (Fryns, 1984, Merenstein et al., 1996).

Similarly, the behavioral phenotypes of FXS can range from mild to more severe levels of intellectual disability. Milder cases can be characterized by a plethora of cognitive and behavioral deficits including hand flapping/biting, hyperactivity, hyperarousal to sensory stimuli, and tactile defensiveness (Merenstein et al., 1996, Hagerman and Hagerman, 2002, Hagerman et al., 2009). Almost all male FXS patients with the full mutation present with some degree of intellectual disability, with an average intelligence quotient (IQ) of around 40 (Rousseau et al., 1991; Hagerman and Hagerman, 2002). Developmental delays, learning disabilities, social phobia, mood disorders, and attention-deficit/hyperactivity disorder can also be present (Merenstein et al., 1996, Sobesky et al., 1996, Franke et al., 1998, Hagerman et al., 1999, Tassone et al., 1999, Tassone et al., 2000, Hagerman and Hagerman, 2002). Comorbidity of FXS with autism spectrum disorders (ASD) occurs in about 30% of cases, making *Fmr1* the largest known single gene cause of autism (Brown et al., 1982, Rogers et al., 2001, Jacquemont et al., 2007).

This book chapter presents an overview of the emerging evidence implicating the hippocampus [a brain region involved in certain aspects of cognition such as learning and memory (Deng et al., 2010) as well as affective behaviors and mood regulation (Ernst et al., 2006; Santarelli et al., 2003)] in the neuropathology of FXS. In particular, this book chapter will discuss the multiple lines of evidence implicating alterations in hippocampal functional (i.e., synaptic) plasticity in the *Fmr1* knock-out (KO) mouse model of FXS (Bakker et al., 1994), focusing on hypofunction of *N*-methyl-D-aspartate receptors (NMDARs) as an emerging aspect in the pathophysiology of FXS. In addition, a discussion on how these changes in hippocampal functional plasticity might mediate alterations in hippocampal-dependent behaviors in *Fmr1* KO mice, and therefore underlie some of the cognitive and behavioral changes seen in FXS, will also be presented.

2. ALTERATIONS IN HIPPOCAMPAL SYNAPTIC PLASTICITY IN *FMR1* KO MICE

2.1. Changes in Short-Term Hippocampal Synaptic Plasticity

In 1994, the Dutch-Belgium Fragile-X Consortium successfully created the *Fmr1* KO mouse by interrupting exon 5 of the *Fmr1* gene with a neomycin cassette (Bakker et al., 1994). Although basal synaptic transmission has been found to be intact in *Fmr1* KO mice, short-term synaptic plasticity was shown to be impaired in the CA1 sub-region of the *Fmr1* hippocampus. An ultrastructural analysis of the CA1 of 2-week old *Fmr1* KO mice demonstrated an immature presynaptic phenotype with smaller synaptic structures, reduced number of vesicles per cluster surface, and an increased fraction of docked vesicles per total number of vesicles (Klemmer et al., 2011). Electrophysiological recordings revealed no significant alterations in CA1 miniature excitatory postsynaptic currents (EPSCs), suggesting no significant changes in synapse strength and glutamate release probability in *Fmr1* KO CA1 pyramidal neurons. However, paired-pulse facilitation and short-term synaptic plasticity were significantly lower in *Fmr1* KO mice when compared to their wild-type controls (Klemmer et al., 2011). In agreement, Deng and colleagues reported that the abnormal short-term synaptic plasticity observed in the CA1 of *Fmr1* KO mice is associated with accelerated vesicle recycling, larger vesicle pools, and increased cumulative calcium influx through N-type and P/Q-type voltage gated calcium channels (Deng et al., 2011). On the other hand, while short-term synaptic depression has been shown to be decreased in amplitude in *Fmr1* KO slices (Deng et al., 2011), short-term potentiation was found to be unaltered (Godfraind et al., 1996).

Idrissi and co-workers have also reported abnormal paired-pulse plasticity in *Fmr1* KO CA1 hippocampal slices, which revealed paired-pulse depression at a comparable inter-stimulus interval that evoked paired-pulse facilitation in slices from wild-type mice (El Idrissi et al., 2010). However, no significant differences in paired-pulse facilitation have been found in other studies (Huber et al., 2002, Larson et al., 2005, Hou et al., 2006, Zhang et al., 2009, Choi et al., 2011, Deng et al., 2011). Differences in the age of the animals at the time of analysis, the background strain in which the mouse colonies were maintained, as well as the stimulation protocols employed to assess paired-pulse facilitation have likely contributed to these discrepancies. Nevertheless, paired-pulse plasticity in *Fmr1* KO slices has been shown to be insensitive to taurine [a γ-aminobutyric acid (GABA)-A receptor agonist] (El Idrissi et al., 2010), possibly due to decreased expression of GABA-A receptors in the hippocampus of the *Fmr1* KO mouse (El Idrissi et al., 2005).

Finally, it is also worth mentioning that loss of presynaptic FMRP can potentially result in a reduction of functional excitatory connections to the CA3 region (Hanson and Madison, 2007). However, paired-pulse plasticity in the DG has been found to be comparable to controls (Eadie et al., 2012; Yun and Trommer, 2011).

2.2. Changes in Long-Term Hippocampal Synaptic Plasticity

The metabotropic glutamate receptor (mGluR)-dependent long-term depression (LTD) theory, in which loss of FMRP leads to alterations in mGluR-triggered synaptic processes (Bear et al., 2004), has inspired a wealth of research in FXS. Huber and colleagues (2012) first reported exaggerated mGluR-dependent LTD in the CA1 hippocampal sub-region with no changes in NMDAR-dependent long-term potentiation (LTP) in the absence of FMRP (Huber et al., 2002). Notably, it has recently been shown that impairments in NMDAR-dependent long-term synaptic plasticity do occur in the DG, but not the CA1 sub-region of the *Fmr1* KO hippocampus (Yun and Trommer, 2011, Franklin et al., 2014; Bostrom et al., 2015). Moreover, therapeutic treatments targeting either inhibition of mGluR (Michalon et al., 2012) or activation of the NMDAR with indirect agonists (Bostrom et al., 2015) can rescue synaptic deficits in *Fmr1* KO mice, further suggesting there are region-specific impairments in glutamatergic signaling in the *Fmr1* KO hippocampus.

Of note, hypofunction of NMDARs is not restricted to the DG of this KO mouse model of FXS and this has also been shown to occur in other brain regions such as the anterior pyriform cortex (Gocel and Larson, 2012), the prefrontal cortex (Martin et al., 2016), and the nucleus accumbens of the striatum (Neuhofer et al., 2015). Therefore, it appears that NMDAR hypofunction may play a prominent role in neuronal dysfunction in several brain regions, thus contributing to various behavioural and cognitive deficits associated with FXS.

2.3. Synaptic Plasticity in the CA1 Sub-Region

It has been proposed that the loss of FMRP may disproportionately impair synaptic plasticity in the CA1 sub-region of *Fmr1* KO mice. Indeed, Bear and colleagues discovered normal low-frequency-induced LTD but exaggerated mGluR5-dependent LTD in the CA1 of juvenile *Fmr1* KO mice (Huber et al., 2002). This finding has been supported by others (Hou et al., 2006, Zhang et al., 2009, Sharma et al., 2010) and recently extended to adult animals (5 – 6 and 9 – 11 months old) (Choi et al., 2011). This increase in mGluR-mediated LTD in the CA1 sub-region of *Fmr1* KO mice has been found to be independent of protein synthesis (Nosyreva and Huber, 2006, Zhang et al., 2009), which is in contrast to what is observed in wild-type animals (Koekkoek et al., 2005, Hou et al., 2006, Nosyreva and Huber, 2006, Zhang et al., 2009). On the other hand, mGluR priming of LTP appears to be similar in both *Fmr1* KO and wild-type CA1 slices, though priming only requires protein synthesis in wild-type slices (Auerbach and Bear, 2010).

Impairments in CA1 LTP have also been found in young *Fmr1* KO mice. Hu and colleagues reported that α-amino-3-hydroxy-5-methyl-4-isoxazolepropionic acid receptors (AMPARs)-mediated synaptic responses are reduced in *Fmr1* KO mice, which may be due to impairment in synaptic delivery of AMPARs containing the GluR1 subunit in this hippocampal sub-region (Hu et al., 2008). However, both LTP induced by high frequency stimulation (Godfraind et al., 1996, Li et al., 2002, Larson et al., 2005) and long-lasting LTP (L-LTP) (Paradee et al., 1999, Zhang et al., 2009) appear to be normal in the CA1 of *Fmr1* KO mice. Theta-burst stimulation (TBS)-induced LTP has also been found to be normal in the *Fmr1* KO CA1 sub-region either when LTP is induced using a single train of 10 theta bursts (Li et al., 2002, Larson et al., 2005, Chen et al., 2010) or 3 trains of 10 bursts (Paradee et al., 1999). The threshold for inducing LTP in the CA1 of *Fmr1* KO mice may be altered however, as 5 trains of TBS were able to unmask LTP deficits (Larson et al., 2005, Lauterborn et al., 2007, Chen et al., 2010, Lee et al., 2011).

Shang et al., reported that basal synaptic transmission and NMDAR-mediated EPSCs were normal in the CA1 sub-region of *Fmr1* KO mice (Shang et al., 2009). In agreement, NMDAR-dependent LTP was also found to be normal in the CA1 of both juvenile and young adult *Fmr1* KO mice (Huber et al., 2002, Eadie et al., 2012, Franklin et al., 2014, Bostrom et al., 2015). However, synaptic LTP induced by glycine (a co-agonist that potentiates the NMDAR) was significantly reduced in the *Fmr1* KO CA1 (Shang et al., 2009).

Of note, long-term synaptic plasticity has also been evaluated in the CA1 sub-region of another FXS mouse model, the *Fmr1* KO2 mouse (Pilpel et al., 2009). Interestingly, while no changes in LTD levels were observed, an enhanced NMDAR-dependent LTP was found in the CA1 sub-region of these KO mice at 2 weeks of age, but not at 6-7 weeks of age. This increase in LTP was associated with a decreased AMPAR/NMDAR ratio, suggesting a down-regulation of AMPAR and up-regulation of NMDAR components (Pilpel et al., 2009).

2.4. Synaptic Plasticity in the DG Sub-Region

The DG is a sub-region of the hippocampus that shows robust alterations in synaptic plasticity in the absence of FMRP. In the *Fmr1* KO mouse, low frequency-induced NMDAR-dependent LTD is attenuated (Eadie et al., 2012). Classical high-frequency-induced LTP has also been found to be impaired in this sub-region (Bostrom et al., 2015; Eadie et al.,

2012; Franklin et al., 2014). Whole cell patch clamping has demonstrated hypofunctionality of NMDARs, which is correlated with an impairment of NMDAR-dependent long-term synaptic plasticity in this brain region (Eadie et al., 2012; Yun and Trommer, 2011). In the *Fmr1* KO mouse, this deficit in NMDAR-dependent long-term synaptic plasticity, specific to the DG sub-region of the hippocampus, has been found to correlate with a down-regulation of NMDAR subunit levels (Bostrom et al., 2015), suggesting that alterations in NMDAR subunit expression may contribute to impaired synaptic plasticity in the DG in the absence of FMRP.

These findings have prompted the recent investigation of pharmacological treatments for impaired NMDAR-dependent synaptic plasticity in the DG of the *Fmr1* KO mouse. The NMDAR co-agonists glycine and D-serine have been shown to potentiate the function of NMDARs (Johnson and Ascher, 1987, Martina et al., 2003). In agreement, bath perfusion of glycine or D-serine were shown to rescue the impairment in NMDAR-dependent LTP in the DG of the *Fmr1* KO mouse, suggesting that promoting NMDAR function may be of therapeutic potential in FXS (Bostrom et al., 2015). Rescue of NMDAR dependent LTP in *Fmr1* KO mice has also been shown through administration of RU38486, a glucocorticoid receptor antagonist (Ghilan et al., 2015). As glucocorticoid receptors can modulate the function and expression of NMDARs (Wang et al., 2005) and the rescue of LTP found by Ghilan et al., was dependent on the NMDAR, this finding may point to abnormal glucocorticoid receptor activity underlying NMDAR hypofunctionality in *Fmr1* KO mice (Ghilan et al., 2015). Another recent finding by Franklin et al., (2014) revealed a relationship between up-regulation of glycogen synthase kinase 3β (GSK3β) activity and impaired NMDAR-dependent LTP in the DG of the *Fmr1* KO mouse. As GSK3β has been shown to impair NMDAR-dependent LTP (Zhu et al., 2007), Franklin and colleagues used pharmacological inhibition of GSK3β to rescue the NMDAR-dependent impairment in LTP observed in *Fmr1* KO mice, which also resulted in improved performance on DG-dependent cognitive tasks (Franklin et al., 2014).

2.5. Relationship between NMDAR- and mGluR-Mediated Synaptic Plasticity in the Hippocampus

Since hypofunction of NMDARs and hyperfunction of mGluR5 are found within different sub-regions of the *Fmr1* KO hippocampus, modulation of these two receptors remains a major focus in FXS research. However, it is of note that mGluRs and NMDARs have the ability to modulate each other, thus regulating their contributions to synaptic plasticity (Harvey et al., 1996, Alagarsamy et al., 1999, Alagarsamy et al., 2002, Alagarsamy et al., 2005, Gerber et al., 2007, Perroy et al., 2008, Bertaso et al., 2010). Applying an mGluR agonist can lead to an enhancement of NMDAR-mediated currents evoked by direct application of NMDA (Harvey and Collingridge, 1993, Awad et al., 2000, Heidinger et al., 2002, Ghilan et al., 2015) and a depression of NMDAR-mediated EPSCs evoked by synaptic stimulation (Baskys and Malenka, 1991, Watabe et al., 2002, Ireland and Abraham, 2009, Ghilan et al., 2015). Whether NMDAR responses are evoked through NMDA application or synaptic stimulation, application of (S)-3,5-dihydroxyphenylglycine (DHPG; a potent agonist of group I mGluRs, including mGluR1 and mGluR5) is known to depress NMDAR responses, possibly through NMDAR internalization (Ghilan et al., 2015). Indeed, mGluR5 may be the mGluR subtype responsible for altering NMDAR responses (Mannaioni et al., 2001, Kotecha et al., 2003, Kotecha and MacDonald, 2003). Furthermore, mGluR5 was shown to co-localize and co-immunoprecipitate with the NMDAR subunit GluN2B (Guo et al., 2004). Of note, type 1 mGluRs may also have the ability to regulate the phosphorylation of both GluN2A and GluN2B NMDAR subunits (Heidinger et al., 2002, Guo et al., 2004).

The likelihood of inducing mGluR-LTD and the magnitude of DHPG-induced mGluR-LTD is known to increase by removal of inhibitory input or removal of Mg^{2+} (which blocks the NMDAR channel) (Harvey et al., 1996, Palmer et al., 1997). Low levels of NMDAR activation can potentiate mGluR5 function while high levels of NMDAR activation can inhibit mGluR5 activation (Alagarsamy et al., 1999, Alagarsamy et al.,

2002). Huang and Hsu (2012) have found that NMDAR antagonists promote mGluR5-LTD induction while activation of NMDARs reduces mGluR5-LTD in the mouse nucleus accumbens. The inhibitory effect of the NMDAR on mGluR5-LTD may be mediated by specific activation of GluN2B-containing NMDARs, since this effect can be blocked by the GluN2B antagonists Ro 25-6981 and ifenprodil, at least in the nucleus accumbens (Huang and Hsu, 2012). These authors suggest that activation of NMDARs results in a calcium/calmodulin-dependent protein kinase II (CaMKII)-mediated increase in phosphorylation of the scaffold protein Homer1b/c, impairing the interaction between mGluR5 and Homer1b/c, thereby attenuating mGluR1/5-LTD induction in this brain region (Huang and Hsu, 2012). On the other hand, mGluR5 in the *Fmr1* hippocampus appears to be less associated with Homer1b/c (Giuffrida et al., 2005), which is possibly related to hyperphosphorylation of Homer1b/c by CamKII as seen in the *Fmr1* cortex (Guo et al., 2015). Since mGluR-LTD is abnormally enhanced in the hippocampus of *Fmr1* KO mice (Huber et al., 2002), the interaction between mGluR5 and Homer1b/c does not appear to be directly related to the amount of mGluR-LTD in the hippocampus of *Fmr1* KO mice. This controversy further highlights the fact that the relationship between mGluRs and NMDARs in the hippocampus of *Fmr1* KO mice is still not completely understood.

Nevertheless, although the interaction between mGluRs and NMDARs in the hippocampus of *Fmr1* KO mice is still unexplored, we speculate that synergistic effects could be observed if mGluR5 inhibition occurs in concert with NMDAR activation. Such pharmacological treatments with the potential to simultaneously increase NMDAR function while also inhibiting mGluR function in the FXS hippocampus (e.g., a cocktail of NMDAR agonists and mGluR antagonists) may prove fruitful in optimizing therapeutic treatments for FXS patients with cognitive impairment. Future studies are warranted in order to test this hypothesis.

3. Alterations in Hippocampal-Dependent Behaviours in *Fmr1* KO Mice

Behavioral deficits in the *Fmr1* KO mice were shown to parallel some of the clinical symptoms manifested by FXS patients. In particular, *Fmr1* KO mice show clear alterations in behaviors that require proper functioning of the hippocampus. These include changes in basal and social anxiety levels, context-dependent learning and memory, fear conditioning, passive avoidance tasks, and subtle alterations in spatial learning. Particularly, behavioral deficits in hippocampal-dependent learning and memory tasks are likely linked to both impaired NMDAR-LTP/LTD in the DG and exaggerated mGluR-LTD in the CA1 sub-region. Indeed, *Fmr1* KO mice display significant impairment in the temporal order task (King and Jope, 2013, Franklin et al., 2014), which requires an intact CA1 sub-region (Hoge and Kesner, 2007), as well as in categorical spatial memory processing, which is dependent on the DG sub-region (King and Jope, 2013, Franklin et al., 2014).

Inhibiting mGluR5 using its antagonist 2-methyl-6-(phenylethynyl)-pyridine (MPEP) has been shown to correct the exaggerated mGluR-LTD in the CA1 region of *Fmr1* KO mice (Dolen et al., 2007, Michalon et al., 2014). In addition, a number of studies have shown that mGluR5 pharmacological blockade (Yan et al., 2005, de Vrij et al., 2008, Gandhi et al., 2014) or reduced mGluR5 expression (Dolen et al., 2007) can rescue several typical FXS phenotypes in *Fmr1* KO mice, including altered opened field activity (Yan et al., 2005), abnormal prepulse inhibition of startle (de Vrij et al., 2008), as well as deficits in some types of learning and memory (Dolen et al., 2007, Gandhi et al., 2014). On the other hand, inhibiting mGluR5 failed to rescue the impairment in DG NMDAR-LTP and was unable to improve the DG-specific pattern separation tasks in *Fmr1* KO mice (Franklin et al., 2014). In contrast, targeting impaired NMDAR-LTP in the DG of *Fmr1* KO mice by pharmacological blockade of its negative regulator GSK3β is able to restore NMDAR-LTP and improve learning and memory tasks including temporal ordering (CA1 dependent), as well as coordinate spatial processing and categorical spatial processing (DG dependent) tasks (Franklin et al., 2014). Taken together, these data support an essential role of NMDARs in hippocampal dysfunction in FXS, and further supports the hypothesis that combined therapies targeting both mGluR5 and NMDARs may have a functional impact in FXS by improving some of the FXS-related behavioral abnormalities.

While the *Fmr1* KO mouse does not exhibit alterations in depression-like behaviors in stressful conditions, as measured through immobility time in the tail suspension and the forced swim tests (Yuskaitis et al., 2010), it does exhibit abnormal anxiety responses. In the open field test, the *Fmr1* KO mouse shows decreased anxiety with increased exploratory activity in the center of the open field (Peier et al., 2000, Qin et al., 2002, Spencer et al., 2005, Eadie et al., 2009, Min et al., 2009, Yuskaitis et al., 2010, Oddi et al., 2015). These KO mice also show an increased propensity of entering the open arms of an elevated plus maze (Mineur et al., 2002, Qin and Smith, 2008, Eadie et al., 2009, Yuskaitis et al., 2010) or elevated zero maze (Bilousova et al., 2009, Liu and Smith, 2009) and spend a decreased amount of time in the dark side of a light-dark chamber with an increased number of transitions between chambers (Peier et al., 2000). Conversely, *Fmr1* KO mice show an increased anxiety response in social situations. They show a greater aversion to going into a mirrored central chamber (Spencer et al., 2005) and display a decreased tendency to win social dominance matches (Spencer et al., 2005). Social defeat causes structural changes in the hippocampus and is functionally important for stress-induced social avoidance (Lagace et al., 2010). The *Fmr1* KO mouse also exhibits other abnormal social interactions in response to direct contact with other mice, such as decreased occurrence of social approach behaviors, abnormal exploration of unfamiliar mice, and lower levels of response to social novelty (Spencer et al., 2005, Liu and Smith, 2009). Passive avoidance is another task that is linked to hippocampal functioning (Isaacson and Wickelgren, 1962, Lorenzini et al., 1996) and was shown to be affected in *Fmr1* KO mice. These FXS mice exhibit an increased probability of entering the side of a chamber where they will be shocked

(Bakker et al., 1994; Qin et al., 2002, Liu and Smith, 2009, Yuskaitis et al., 2010). Another behavioral task that requires the hippocampus is contextual and trace fear conditioning (Phillips and LeDoux, 1992, McEchron et al., 1998). Whereas a few studies have reported decreased freezing in contextual fear conditioning in *Fmr1* KO mice (Paradee et al., 1999, Busquets-Garcia et al., 2013, Oddi et al., 2015), others were unable to replicate these results (Dobkin et al., 2000, Peier et al., 2000, Van Dam et al., 2000). However, since contextual fear is also dependent on proper amygdala functioning, the interpretation of these findings in the context of *Fmr1* KO hippocampal function may not be straightforward (Paradee et al., 1999).

Other behaviors that are dependent on hippocampal function include spatial learning and memory. Although *Fmr1* KO mice show deficits in spatial learning and memory as tested using the radial maze (Mineur et al., 2002) and the plus-shaped water maze (Van Dam et al., 2000), conflicting results have been reported regarding the performance of these KO mice in the Morris Water Maze (MWM) test, one of the behavioral tests most used to assess hippocampal-dependent spatial learning and memory (Morris, 1984). Initial studies found that *Fmr1* KO mice showed spatial learning and memory deficits in the MWM, including deficits in the acquisition of learning (Kooy et al., 1996), reversal trials (Bakker et al., 1994, Kooy et al., 1996, D'Hooge et al., 1997, Gantois et al., 2001), and possible motor impairments (D'Hooge et al., 1997). However, others were unable to find robust MWM learning deficits in these FXS mice (Paradee et al., 1999, Peier et al., 2000). Differences in the genetic background of the mice used in the various studies may underlie these inconsistencies in MWM performance (Simpson et al., 1997, Threadgill et al., 1997, Paradee et al., 1999, Dobkin et al., 2000). Another learning and memory task dependent on hippocampal functioning is context discrimination. *Fmr1* KO mice show an impairment in context discrimination, with an increased rate of freezing in a context not paired with a shock on day 7 of a contextual fear extinction assay (Eadie et al., 2012). Interestingly, this contextual discrimination task is normally dependent on functional NMDARs in the DG (McHugh et al., 2007), which provides a clear behavioral consequence of the deficits in NMDAR function and long-term synaptic plasticity seen in this sub-region of the *Fmr1* KO hippocampus (Yun and Trommer, 2011, Eadie et al., 2012, Franklin et al., 2014, Bostrom et al., 2015).

Conclusion

Research conducted over the last 25 years has determined that the absence of FMRP affects several physiological systems, including multiple pathways within the nervous system. The hippocampus is one brain region that repeatedly appears to be particularly affected by the loss of FMRP, both structurally and functionally, and thus is thought to contribute to cognitive impairment in this neurodevelopmental disorder.

Previous research has focused on the mGluR theory of FXS with evidence showing that the loss of FMRP adversely affects synaptic plasticity in the CA1 hippocampal sub-region. Emerging findings on NMDAR hypofunction in the DG sub-region of the hippocampus have unravelled the potential role of NMDAR in the impairment of hippocampal dependent learning and memory observed in FXS. *In vivo* studies using *Fmr1* KO mice have demonstrated that treatments restoring NMDAR-mediated plasticity also rescue some types of hippocampal dependent learning and memory. These results highlight the fact that increasing hippocampal NMDAR function (alone or in combination with decreasing hippocampal mGluR5 activity) may have therapeutic potential by improving some aspects of cognitive functioning in FXS patients.

However, clinical trials aimed at reversing NMDAR hypofunction in FXS have not been attempted to date, though recent findings suggest that indirectly facilitating NMDAR function may have the potential to rescue cognitive functioning in FXS. Promoting NMDAR functioning may be a future therapeutic option for FXS individuals with cognitive impairment. Furthermore, combination therapies that can simultaneously target mGluRs and NMDARs in the hippocampus may show cumulative beneficial effects on improving hippocampal dysfunction in FXS patients with cognitive impairment.

ACKNOWLEDGMENTS

JGM acknowledges research funding from the *Ciência Sem Fronteiras/CNPq* (Science Without Borders) funding program of the Brazilian Federal Government.

REFERENCES

Alagarsamy S, Marino MJ, Rouse ST, Gereau RWT, Heinemann SF, Conn PJ (1999) Activation of NMDA receptors reverses desensitization of mGluR5 in native and recombinant systems. *Nat Neurosci* 2: 234-240.

Alagarsamy S, Rouse ST, Junge C, Hubert GW, Gutman D, Smith Y, Conn PJ (2002) NMDA-induced phosphorylation and regulation of mGluR5. *Pharmacol Biochem Behav* 73: 299-306.

Alagarsamy S, Saugstad J, Warren L, Mansuy IM, Gereau RWT, Conn PJ (2005) NMDA-induced potentiation of mGluR5 is mediated by activation of protein phosphatase 2B/calcineurin. *Neuropharmacology* 1: 135-145.

Antar LN, Bassell GJ (2003) Sunrise at the synapse: the FMRP mRNP shaping the synaptic interface. *Neuron* 37: 555-558.

Ashley CT, Sutcliffe JS, Kunst CB, Leiner HA, Eichler EE, Nelson DL, Warren ST (1993) Human and murine FMR-1: alternative splicing and translational initiation downstream of the CGG-repeat. *Nat Genet* 4: 244-251.

Auerbach BD, Bear MF (2010) Loss of the fragile X mental retardation protein decouples metabotropic glutamate receptor dependent priming of long-term potentiation from protein synthesis. *J Neurophysiol* 104: 1047-1051.

Awad H, Hubert GW, Smith Y, Levey AI, Conn PJ (2000) Activation of metabotropic glutamate receptor 5 has direct excitatory effects and potentiates NMDA receptor currents in neurons of the subthalamic nucleus. *J Neurosci* 20: 7871-7879.

Bakker CE, Verheij C, Willemsen R, Van Der Helm R, Oerlemans F, Vermey M, Bygrave A, Hoogeveen A, Oostra BA, Reynierse E, De Boulee K, D'Hoogeg R, Crasf P, Van Velzene D, Nagelsg G, Martinf J, De Deyng PP, Darbye JK, Willems PJ (1994) Fmr1 knockout mice: a model to study fragile X mental retardation. The Dutch-Belgian Fragile X Consortium. *Cell* 78: 23-33.

Baskys A, Malenka RC (1991) Agonists at metabotropic glutamate receptors presynaptically inhibit EPSCs in neonatal rat hippocampus. *J Physiol* 444: 687-701.

Bassell GJ, Warren ST (2008) Fragile X syndrome: loss of local mRNA regulation alters synaptic development and function. *Neuron* 60: 201-214.

Bear MF, Huber KM, Warren ST (2004) The mGluR theory of fragile X mental retardation. *Trend Neurosci* 27: 370-377.

Bertaso F, Roussignol G, Worley P, Bockaert J, Fagni L, Ango F (2010) Homer1a-dependent crosstalk between NMDA and metabotropic glutamate receptors in mouse neurons. *PLoS One* 5: e9755.

Bilousova TV, Dansie L, Ngo M, Aye J, Charles JR, Ethell DW, Ethell IM (2009) Minocycline promotes dendritic spine maturation and improves behavioural performance in the fragile X mouse model. *J Med Genet* 46: 94-102.

Bostrom CA, Majaess NM, Morch K, White E, Eadie BD, Christie BR (2015) Rescue of NMDAR-dependent synaptic plasticity in Fmr1 knock-out mice. *Cereb Cortex* 25: 271-279.

Brown WT, Friedman E, Jenkins EC, Brooks J, Wisniewski K, Raguthu S, French JH (1982) Association of fragile X syndrome with autism. *Lancet* 1: 100.

Busquets-Garcia A, Gomis-Gonzalez M, Guegan T, Agustin-Pavon C, Pastor A, Mato S, Perez-Samartin A, Matute C, de la Torre R., Dierssen M, Maldonado R, Ozaita A (2013) Targeting the endocannabinoid system in the treatment of fragile X syndrome. *Nat Med* 19: 603-607.

Chen LY, Rex CS, Babayan AH, Kramar EA, Lynch G, Gall CM, Lauterborn JC (2010) Physiological activation of synaptic Rac > PAK (p-21 activated kinase) signaling is defective in a mouse model of fragile X syndrome. *J Neurosci* 30: 10977-10984.

Choi CH, Schoenfeld BP, Bell AJ, Hinchey P, Kollaros M, Gertner MJ, Woo NH, Tranfaglia MR, Bear MF, Zukin RS, McDonald TV, Jongens TA, McBride SM (2011) Pharmacological reversal of synaptic plasticity deficits in the mouse model of fragile X syndrome by group II mGluR antagonist or lithium treatment. *Brain Res*. 1380: 106-119.

Christie BR, Cameron HA (2006) Neurogenesis in the adult hippocampus. *Hippocampus* 16: 199-207.

Chudley AE, Hagerman RJ (1987) Fragile X syndrome. *J Pediatr* 110: 821-831.

D'Hooge R, Nagels G, Franck F, Bakker CE, Reyniers E, Storm K, Kooy RF, Oostra BA, Willems PJ, De Deyn PP (1997) Mildly impaired water maze performance in male Fmr1 knockout mice. *Neurosci* 76: 367-376.

de Vrij FM, Levenga J, van der Linde HC, Koekkoek SK, De Zeeuw CI, Nelson DL, Oostra BA, Willemsen R (2008) Rescue of behavioral phenotype and neuronal protrusion morphology in Fmr1 KO mice. *Neurobiol Dis* 31: 127-132.

Deng PY, Sojka D, Klyachko VA (2011) Abnormal presynaptic short-term plasticity and information processing in a mouse model of fragile X syndrome. *J Neurosci* 31: 10971-10982.

Deng W, Aimone JB, Gage FH (2010) New neurons and new memories: how does adult hippocampal neurogenesis affect learning and memory? *Nat Rev Neurosci* 11: 339-350.

Dobkin C, Rabe A, Dumas R, El Idrissi A, Haubenstock H, Brown WT (2000) Fmr1 knockout mouse has a distinctive strain-specific learning impairment. *Neurosci* 100: 423-429.

Dolen G, Osterweil E, Rao BS, Smith GB, Auerbach BD, Chattarji S, Bear MF (2007) Correction of fragile X syndrome in mice. *Neuron* 56: 955-962.

Eadie BD, Cushman J, Kannangara TS, Fanselow MS, Christie BR (2012) NMDA receptor hypofunction in the dentate gyrus and impaired context discrimination in adult Fmr1 knockout mice. *Hippocampus* 22: 241-254.

Eadie BD, Zhang WN, Boehme F, Gil-Mohapel J, Kainer L, Simpson JM, Christie BR (2009) Fmr1 knockout mice show reduced anxiety and alterations in neurogenesis that are specific to the ventral dentate gyrus. *Neurobiol Dis.* 36: 361-373.

El Idrissi A, Ding XH, Scalia J, Trenkner E, Brown WT, Dobkin C (2005) Decreased GABA(A) receptor expression in the seizure-prone fragile X mouse. *Neurosci Lett* 377: 141-146.

El Idrissi A, Neuwirth LS, L'Amoreaux W (2010) Taurine regulation of short term synaptic plasticity in fragile X mice. *J Biomed Sci* 17 (Suppl 1): S15.

Ernst C, Olson AK, Pinel JP, Lam RW, Christie BR (2006) Antidepressant effects of exercise: evidence for an adult-neurogenesis hypothesis? *J Psychiatry Neurosci* 31: 84-92.

Franke P, Leboyer M, Gansicke M, Weiffenbach O, Biancalana V, Cornillet-Lefebre P, Croquette MF, Froster U, Schwab SG, Poustka F, Hautzinger M, Maier W (1998) Genotype-phenotype relationship in female carriers of the premutation and full mutation of FMR-1. *Psychiatry Res* 80: 113-127.

Franklin AV, King MK, Palomo V, Martinez A, McMahon LL, Jope RS (2014) Glycogen synthase kinase-3 inhibitors reverse deficits in long-term potentiation and cognition in fragile X mice. *Biol Psychiatry* 75: 198-206.

Fryns JP (1984) The fragile X syndrome. A study of 83 families. *Clin Genet* 26: 497-528.

Fu YH, Kuhl DP, Pizzuti A, Pieretti M, Sutcliffe JS, Richards S, Verkerk AJ, Holden JJ, Fenwick RG, Jr, Warren ST, Oostra BA, Nelson DL, Caskey CT (1991) Variation of the CGG repeat at the fragile X site results in genetic instability: resolution of the Sherman paradox. *Cell.* 67: 1047-1058.

Gandhi RM, Kogan CS, Messier C (2014) 2-Methyl-6-(phenylethynyl) pyridine (MPEP) reverses maze learning and PSD-95 deficits in Fmr1 knock-out mice. *Front Cell Neurosci* 8: 70.

Gantois I, Bakker CE, Reyniers E, Willemsen R, D'Hooge R, De Deyn PP, Oostra BA, Kooy RF (2001) Restoring the phenotype of fragile X syndrome: insight from the mouse model. *Curr Mol Med.* 1, 447-455.

Gerber U, Gee CE, Benquet P (2007) Metabotropic glutamate receptors: intracellular signaling pathways. *Curr Opin Pharmacol* 7: 56-61.

Ghilan M, Hryciw BN, Brocardo PS, Bostrom CA, Gil-Mohapel J, Christie BR (2015) Enhanced corticosteroid signaling alters synaptic plasticity in the dentate gyrus in mice lacking the fragile X mental retardation protein. *Neurobiol Dis* 77: 26-34.

Giuffrida R, Musumeci S, D'Antoni S, Bonaccorso CM, Giuffrida-Stella AM, Oostra BA, Catania MV (2005) A reduced number of metabotropic glutamate subtype 5 receptors are associated with constitutive homer proteins in a mouse model of fragile X syndrome. *J Neurosci* 28: 8908-8916.

Gocel J, Larson J (2012) Synaptic NMDA receptor-mediated currents in anterior piriform cortex are reduced in the adult fragile X mouse. *Neurosci* 221: 170-181.

Godfraind JM, Reyniers E, De Boulle K, D'Hooge R, De Deyn PP, Bakker CE, Oostra BA, Kooy RF, Willems PJ (1996) Long-term potentiation in the hippocampus of fragile X knockout mice. *Am J Med Genet* 64: 246-251.

Guo W, Ceolin L, Collins KA, Perroy J, Huber KM (2015) Elevated CaMKIIalpha and Hyperphosphorylation of Homer Mediate Circuit Dysfunction in a Fragile X Syndrome Mouse Model *Cell Rep* 13: 2297-2311.

Guo W, Wei F, Zou S, Robbins MT, Sugiyo S, Ikeda T, Tu JC, Worley PF, Dubner R, Ren K (2004) Group I metabotropic glutamate receptor NMDA receptor coupling and signaling cascade mediate spinal dorsal horn NMDA receptor 2B tyrosine phosphorylation associated with inflammatory hyperalgesia. *J Neurosci* 24: 9161-9173.

Hagerman R, Van Housen K, Smith A, McGavran L (1984) Consideration of connective tissue dysfunction in the fragile X syndrome. *Am J Med Genet* 17: 111-121.

Hagerman RJ, Berry-Kravis E, Kaufmann WE, Ono MY, Tartaglia N, Lachiewicz A, Kronk R, Delahunty C, Hessl D, Visootsak J, Picker J, Gane L, Tranfaglia M (2009) Advances in the treatment of fragile X syndrome. *Pediatrics* 123: 378-390.

Hagerman RJ, Hagerman PJ (2002) *Fragile X Syndrome: Diagnosis, Treatment, and Research:* Johns Hopkins University Press.

Hagerman RJ, Hills J, Scharfenaker S, Lewis H (1999) Fragile X syndrome and selective mutism. *Am J Med Genet* 83: 313-317.

Hanson JE, Madison DV (2007) Presynaptic FMR1 genotype influences the degree of synaptic connectivity in a mosaic mouse model of fragile X syndrome. *J Neurosci* 27: 4014-4018.

Harvey J, Collingridge GL (1993) Signal transduction pathways involved in the acute potentiation of NMDA responses by 1S, 3R-ACPD in rat hippocampal slices. *Br J Pharmacol* 109: 1085-1090.

Harvey J, Palmer MJ, Irving AJ, Clarke VR, Collingridge GL (1996) NMDA receptor dependence of mGlu-mediated depression of synaptic transmission in the CA1 region of the rat hippocampus. *Br J Pharmacol* 119: 1239-1247.

Heidinger V, Manzerra P, Wang XQ, Strasser U, Yu SP, Choi DW, Behrens MM (2002) Metabotropic glutamate receptor 1-induced upregulation of NMDA receptor current: mediation through the Pyk2/Src-family kinase pathway in cortical neurons. *J Neurosci* 22: 5452-5461.

Hoge J, Kesner RP (2007) Role of CA3 and CA1 subregions of the dorsal hippocampus on temporal processing of objects. *Neurobiol Learn Mem* 88: 225-231.

Hornstra IK, Nelson DL, Warren ST, Yang TP (1993) High resolution methylation analysis of the FMR1 gene trinucleotide repeat region in fragile X syndrome. *Hum Mol Genet* 2: 1659-1665.

Hou L, Antion MD, Hu D, Spencer CM, Paylor R, Klann E (2006) Dynamic translational and proteasomal regulation of fragile X mental retardation protein controls mGluR-dependent long-term depression. *Neuron* 51: 441-454.

Hu H, Qin Y, Bochorishvili G, Zhu Y, van Aelst L, Zhu JJ (2008) Ras signaling mechanisms underlying impaired GluR1-dependent plasticity associated with fragile X syndrome. *J Neurosci* 28: 7847-7862.

Huang CC, Hsu KS (2012) Activation of NMDA receptors reduces metabotropic glutamate receptor-induced long-term depression in the nucleus accumbens via a CaMKII-dependent mechanism. *Neuropharmacology* 63: 1298-1307.

Huber KM, Gallagher SM, Warren ST, Bear MF (2002) Altered synaptic plasticity in a mouse model of fragile X mental retardation. *Proc Natl Acad Sci US* 99: 7746-7750.

Ireland DR, Abraham WC (2009) Mechanisms of group I mGluR-dependent long-term depression of NMDA receptor-mediated transmission at Schaffer collateral-CA1 synapses. *J Neurophysiol* 101: 1375-1385.

Isaacson RL, Wickelgren WO (1962) Hippocampal ablation and passive avoidance. *Science* 138: 1104-1106.

Jacquemont S, Hagerman RJ, Hagerman PJ, Leehey MA (2007) Fragile-X syndrome and fragile X-associated tremor/ataxia syndrome: two faces of FMR1. *Lancet Neurol* 6: 45-55.

Johnson JW, Ascher P (1987) Glycine potentiates the NMDA response in cultured mouse brain neurons. *Nature* 325: 529-531.

King MK, Jope RS (2013) Lithium treatment alleviates impaired cognition in a mouse model of fragile X syndrome. *Genes Brain Behav* 12: 723-731.

Klemmer P, Meredith RM, Holmgren CD, Klychnikov OI, Stahl-Zeng J, Loos M, van der Schors RC, Wortel J, de Wit H, Spijker S, Rotaru DC, Mansvelder HD, Smit AB, Li KW (2011) Proteomics, ultrastructure, and physiology of hippocampal synapses in a fragile X syndrome mouse model reveal presynaptic phenotype. *J Biol Chem* 286: 25495-25504.

Koekkoek SK, Yamaguchi K, Milojkovic BA, Dortland BR, Ruigrok TJ, Maex R, De Graaf W, Smit AE, VanderWerf F, Bakker CE, Willemsen R, Ikeda T, Kakizawa S, Onodera K, Nelson DL, Mientjes E, Joosten M, De Schutter E, Oostra BA, Ito M, De Zeeuw CI (2005)

Deletion of FMR1 in Purkinje cells enhances parallel fiber LTD, enlarges spines, and attenuates cerebellar eyelid conditioning in Fragile X syndrome. *Neuron* 47: 339-352.

Kooy RF, D'Hooge R, Reyniers E, Bakker CE, Nagels G, De Boulle K, Storm K, Clincke G, De Deyn PP, Oostra BA, Willems PJ (1996) Transgenic mouse model for the fragile X syndrome. *Am J Med Genet* 64: 241-245.

Kotecha SA, Jackson MF, Al-Mahrouki A, Roder JC, Orser BA, MacDonald JF (2003) Co-stimulation of mGluR5 and N-methyl-D-aspartate receptors is required for potentiation of excitatory synaptic transmission in hippocampal neurons. *J Biol Chem* 278: 27742-27749.

Kotecha SA, MacDonald JF (2003) Signaling molecules and receptor transduction cascades that regulate NMDA receptor-mediated synaptic transmission. *Int Rev Neurobiol* 54: 51-106.

Lagace DC, Donovan MH, DeCarolis NA, Farnbauch LA, Malhotra S, Berton O, Nestler EJ, Krishnan V, Eisch AJ (2010) Adult hippocampal neurogenesis is functionally important for stress-induced social avoidance. *Proc Natl Acad Sci US* 107: 4436-4441.

Larson J, Jessen RE, Kim D, Fine AK, du Hoffmann J (2005) Age-dependent and selective impairment of long-term potentiation in the anterior piriform cortex of mice lacking the fragile X mental retardation protein. *J Neurosci* 25: 9460-9469.

Lauterborn JC, Rex CS, Kramar E, Chen LY, Pandyarajan V, Lynch G, Gall CM (2007) Brain-derived neurotrophic factor rescues synaptic plasticity in a mouse model of fragile X syndrome. *J Neurosci* 27: 10685-10694.

Lee HY, Ge WP, Huang W, He Y, Wang GX, Rowson-Baldwin A, Smith SJ, Jan YN, Jan LY (2011) Bidirectional regulation of dendritic voltage-gated potassium channels by the fragile X mental retardation protein. *Neuron* 72: 630-642.

Li J, Pelletier MR, Perez Velazquez JL, Carlen PL (2002) Reduced cortical synaptic plasticity and GluR1 expression associated with fragile X mental retardation protein deficiency. *Mol Cell Neurosci* 19: 138-151.

Liu ZH, Smith CB (2009) Dissociation of social and nonsocial anxiety in a mouse model of fragile X syndrome. *Neurosci Lett* 454: 62-66.

Lorenzini CA, Baldi E, Bucherelli C, Sacchetti B, Tassoni G (1996) Role of dorsal hippocampus in acquisition, consolidation and retrieval of rat's passive avoidance response: a tetrodotoxin functional inactivation study. *Brain Res* 730: 32-39.

Mannaioni G, Marino MJ, Valenti O, Traynelis SF, Conn PJ (2001) Metabotropic glutamate receptors 1 and 5 differentially regulate CA1 pyramidal cell function. *J Neurosci* 21: 5925-5934.

Martin HG, Lassalle O, Brown JT, Manzoni OJ (2016) Age-Dependent Long-Term Potentiation Deficits in the Prefrontal Cortex of the Fmr1 Knockout Mouse Model of Fragile X Syndrome. *Cereb Cortex* 26: 2084-2092.

Martina M, Krasteniakov NV, Bergeron R (2003) D-Serine differently modulates NMDA receptor function in rat CA1 hippocampal pyramidal cells and interneurons. *J Physiol* 548: 411-423.

McEchron MD, Bouwmeester H, Tseng W, Weiss C, Disterhoft JF (1998) Hippocampectomy disrupts auditory trace fear conditioning and contextual fear conditioning in the rat. *Hippocampus*. 8: 638-646.

McHugh TJ, Jones MW, Quinn JJ, Balthasar N, Coppari R, Elmquist JK, Lowell BB, Fanselow MS, Wilson MA, Tonegawa S (2007) Dentate gyrus NMDA receptors mediate rapid pattern separation in the hippocampal network. *Science* 317: 94-99.

Merenstein SA, Sobesky WE, Taylor AK, Riddle JE, Tran HX, Hagerman RJ (1996) Molecular-clinical correlations in males with an expanded FMR1 mutation. *Am J Med Genet* 64: 388-394.

Michalon A, Bruns A, Risterucci C, Honer M, Ballard TM, Ozmen L, Jaeschke G, Wettstein JG, von Kienlin M, Kunnecke B, Lindemann L (2014) Chronic metabotropic glutamate receptor 5 inhibition corrects local alterations of brain activity and improves cognitive performance in fragile X mice. *Biol Psychiatry* 75: 189-197.

Michalon A, Sidorov M, Ballard TM, Ozmen L, Spooren W, Wettstein JG, Jaeschke G, Bear MF, Lindemann L (2012) Chronic pharmacological mGlu5 inhibition corrects fragile X in adult mice. *Neuron* 74: 49-56.

Min WW, Yuskaitis CJ, Yan Q, Sikorski C, Chen S, Jope RS, Bauchwitz RP (2009) Elevated glycogen synthase kinase-3 activity in Fragile X mice: Key metabolic regulator with evidence for treatment potential. *Neuropharmacology* 56: 463-472.

Mineur YS, Sluyter F, de Wit S, Oostra BA, Crusio WE (2002) Behavioral and neuroanatomical characterization of the Fmr1 knockout mouse. *Hippocampus* 12: 39-46.

Morris R (1984) Developments of a water-maze procedure for studying spatial learning in the rat. *J Neurosci Methods* 11: 47-60.

Neuhofer D, Henstridge CM, Dudok B, Sepers M, Lassalle O, Katona I, Manzoni OJ (2015) Functional and structural deficits at accumbens synapses in a mouse model of Fragile X Front *Cell Neurosci* 9: 100.

Nosyreva ED, Huber KM (2006) Metabotropic receptor-dependent long-term depression persists in the absence of protein synthesis in the mouse model of fragile X syndrome. *J Neurophysiol* 95: 3291-3295.

O'Donnell WT, Warren ST (2002) A decade of molecular studies of fragile X syndrome. *Annu Rev Neurosci* 25: 315-338.

Oddi D, Subashi E, Middei S, Bellocchio L, Lemaire-Mayo V, Guzman M, Crusio WE, D'Amato FR, Pietropaolo S (2015) Early social enrichment rescues adult behavioral and brain abnormalities in a mouse model of fragile X syndrome. *Neuropsychopharmacology* 40: 1113-1122.

Palmer MJ, Irving AJ, Seabrook GR, Jane DE, Collingridge GL (1997) The group I mGlu receptor agonist DHPG induces a novel form of LTD in the CA1 region of the hippocampus. *Neuropharmacology* 36: 1517-1532.

Paradee W, Melikian HE, Rasmussen DL, Kenneson A, Conn PJ, Warren ST (1999) Fragile X mouse: strain effects of knockout phenotype and evidence suggesting deficient amygdala function. *Neuroscience* 94: 185-192.

Partington MW (1984) The fragile X syndrome II: preliminary data on growth and development in males. *Am J Med Genet* 17: 175-194.

Peier AM, McIlwain KL, Kenneson A, Warren ST, Paylor R, Nelson DL (2000) (Over)correction of FMR1 deficiency with YAC transgenics: behavioral and physical features. *Hum Mol Genet* 9: 1145-1159.

Perroy J, Raynaud F, Homburger V, Rousset MC, Telley L, Bockaert J, Fagni L (2008) Direct interaction enables cross-talk between ionotropic and group I metabotropic glutamate receptors. *J Biol Chem* 283: 6799-6805.

Phillips RG, LeDoux JE (1992) Differential contribution of amygdala and hippocampus to cued and contextual fear conditioning. *Behav Neurosci* 106: 274-285.

Pieretti M, Zhang F, Fu YH, Warren ST, Oostra BA, Caskey CT, Nelson DL (1991) Absence of expression of the FMR-1 gene in fragile X syndrome. *Cell* 66: 817-822.

Pilpel Y, Kolleker A, Berberich S, Ginger M, Frick A, Mientjes E, Oostra BA, Seeburg PH (2009) Synaptic ionotropic glutamate receptors and plasticity are developmentally altered in the CA1 field of Fmr1 knockout mice. *J Physiol* 587: 787-804.

Qin M, Kang J, Smith CB (2002) Increased rates of cerebral glucose metabolism in a mouse model of fragile X mental retardation. *Proc Natl Acad Sci USA* 99: 15758-15763.

Qin M, Smith CB (2008) Unaltered hormonal response to stress in a mouse model of fragile X syndrome. *Psychoneuroendocrinology* 33: 883-889.

Reiss AL, Freund LS, Baumgardner TL, Abrams MT, Denckla MB (1995) Contribution of the FMR1 gene mutation to human intellectual dysfunction. *Nat Genet*. 11: 331-334.

Rogers SJ, Wehner DE, Hagerman R (2001) The behavioral phenotype in fragile X: symptoms of autism in very young children with fragile X syndrome, idiopathic autism, and other developmental disorders. *J Dev Behav Pediatr* 22: 409-417.

Rousseau F, Heitz D, Biancalana V, Blumenfeld S, Kretz C, Boue J, Tommerup N, Van Der Hagen C, DeLozier-Blanchet C, Croquette MF, Gilgenkrantz S, Jalbert P, Voelckel MA, Oberlé I, Mandel JL (1991) Direct diagnosis by DNA analysis of the fragile X syndrome of mental retardation. *N Engl J Med* 325: 1673-1681.

Santarelli L, Saxe M, Gross C, Surget A, Battaglia F, Dulawa S, Weisstaub N, Lee J, Duman R, Arancio O, Belzung C, Hen R (2003) Requirement of hippocampal neurogenesis for the behavioral effects of antidepressants. *Science*. 301: 805-809.

Shang Y, Wang H, Mercaldo V, Li X, Chen T, Zhuo M (2009) Fragile X mental retardation protein is required for chemically-induced long-term potentiation of the hippocampus in adult mice. *J Neurochem* 111: 635-646.

Sharma A, Hoeffer CA, Takayasu Y, Miyawaki T, McBride SM, Klann E, Zukin RS (2010) Dysregulation of mTOR signaling in fragile X syndrome. *J Neurosci* 30: 694-702.

Simpson EM, Linder CC, Sargent EE, Davisson MT, Mobraaten LE, Sharp JJ (1997) Genetic variation among 129 substrains and its importance for targeted mutagenesis in mice. *Nat Genet* 16: 19-27.

Sobesky WE, Taylor AK, Pennington BF, Bennetto L, Porter D, Riddle J, Hagerman RJ (1996) Molecular/clinical correlations in females with fragile X. *Am J Med Genet*: 64, 340-345.

Spencer CM, Alekseyenko O, Serysheva E, Yuva-Paylor LA, Paylor R (2005) Altered anxiety-related and social behaviors in the Fmr1 knockout mouse model of fragile X syndrome. *Genes Brain Behav* 4: 420-430.

Sutcliffe JS, Nelson DL, Zhang F, Pieretti M., Caskey CT, Saxe D., Warren ST, 1992. DNA methylation represses FMR-1 transcription in fragile X syndrome. *Hum Mol Genet*. 1, 397-400.

Tassone F, Hagerman RJ, Ikle DN, Dyer PN, Lampe M, Willemsen R, Oostra BA, Taylor AK (1999) FMRP expression as a potential prognostic indicator in fragile X syndrome. *Am J Med Genet*. 84: 250-261.

Tassone F, Hagerman RJ, Taylor AK, Mills JB, Harris SW, Gane LW, Hagerman PJ (2000) Clinical involvement and protein expression in individuals with the FMR1 premutation. *Am J Med Genet*. 91, 144-152.

Threadgill DW, Matin A, Yee D, Carrasquillo MM., Henry KR, Rollins KG, Nadeau JH, Magnuson T (1997) SSLPs to map genetic differences between the 129 inbred strains and closed-colony, random-bred CD-1 mice. *Mamm Genome* 8: 441-442.

Turner G, Daniel A, Frost M (1980) X-linked mental retardation, macro-orchidism, and the Xq27 fragile site. *The Journal of Pediatrics* 96: 837-841.

Turner G, Webb T, Wake S, Robinson H (1996) Prevalence of fragile X syndrome. *Am J Med Genet* 64: 196-197.

Van Dam D, D'Hooge R, Hauben E, Reyniers E, Gantois I, Bakker CE, Oostra BA, Kooy RF, De Deyn PP (2000) Spatial learning, contextual fear conditioning and conditioned emotional response in Fmr1 knockout mice. *Behav Brain Res* 117: 127-136.

Verma IC, Elango R (1994) Variable expression of clinical features of Martin Bell syndrome in younger patients. *Indian Pediatr* 31: 433-438.

Watabe AM, Carlisle HJ, O'Dell TJ (2002) Postsynaptic induction and presynaptic expression of group 1 mGluR-dependent LTD in the hippocampal CA1 region. *J Neurophysiol* 87: 1395-1403.

Yan QJ, Rammal M, Tranfaglia M, Bauchwitz RP (2005) Suppression of two major Fragile X Syndrome mouse model phenotypes by the mGluR5 antagonist MPEP. N*europharmacology* 49: 1053-1066.

Yun SH, Trommer BL (2011) Fragile X mice: reduced long-term potentiation and N-Methyl-D-Aspartate receptor-mediated neurotransmission in dentate gyrus. *J Neurosci Res* 89: 176-182.

Yuskaitis CJ, Mines MA, King MK, Sweatt JD, Miller CA, Jope RS (2010) Lithium ameliorates altered glycogen synthase kinase-3 and behavior in a mouse model of fragile X syndrome. *Biochem Pharmacol* 79: 632-646.

Zhang J, Hou L, Klann E, Nelson DL (2009) Altered hippocampal synaptic plasticity in the FMR1 gene family knockout mouse models. *J Neurophysiol* 101: 2572-2580.

Zhu LQ, Wang SH, Liu D, Yin YY, Tian Q, Wang XC, Wang Q, Chen JG, Wang JZ (2007) Activation of glycogen synthase kinase-3 inhibits long-term potentiation with synapse-associated impairments. *J Neurosci* 27: 12211-12220.

In: Synaptic Plasticity
Editor: Sraboni Chaudhury
ISBN: 978-1-53613-606-7

Chapter 7

The Role of Neurotrophic Factors in Regulating Synaptic Plasticity in Psychiatric Disorder

Sraboni Chaudhury
Molecular and Behavioral Neuroscience Institute,
University of Michigan, MI, US

Abstract

Neurotrophic factor plays an important role in many brain functions during development as well as in adulthood. Environmental factors are also known to regulate growth factors. Therefore, it is important to understand the interactions between genes and environment that are involved in brain development as well as during various neurological diseases including psychiatric illness. In this perspective, it appears crucial to know the way, an organism respond to environmental stimuli and provide a coordinated and adapted response. In the central nervous system, neuronal plasticity and neurotransmission are among the major processes that are integrating such complex interactions between genes and environmental stimuli. Several growth factors (NGF- Nerve growth factor, BDNF- Brain-derived Neurotrophic factor and Neurotrophins), are critical components of influencing synaptic plasticity which provides the molecular framework for a rapid and dynamic response to neuronal activity. This could lead to a long lasting and sustained adaptation through regulation of number of gene expression in an individual. As a result, these growth factors have been strongly associated with neuronal activity as well as a variety of higher order processes within the central nervous system such as learning, memory and sensitivity to reward. Various growth factors have been revealed as a major mediator and regulator of synaptic plasticity and neuronal activity in both physiological and pathological conditions. In this chapter, we will focus on the role of neurotrophins in the central nervous system and how they influence synaptic plasticity by modifying the environment. Finally, we will discuss and illustrate the role of growth factors in regulating brain plasticity in pathological states with a particular interest in cognitive functions and neuropsychiatric disorders.

Keywords: Neurotrophins, synaptic plasticity, psychiatric disorder, Nerve growth factor, Brain-derived Neurotrophic factor

1. INTRODUCTION

Nerve growth factor (NGF), brain-derived neurotrophic factor (BDNF), neurotrophin-3 (NT-3), and neurotrophin-4/5 (NT-4/5) are derived from a common ancestral gene and are together called as neurotrophins (Hallbook 1999). Neurotrophic factors are a family of highly conserved proteins in mammal. All neurotrophins have a similar overall structure, with a small number of amino acid differences which determines the specificity of receptor binding sites. The mature active forms of neurotrophins are very stable and noncovalently associated homodimers, with molecular weights around 28 kDa. Dimerization seems to be an essential requisite and characteristic feature for all neurotrophins which is vital for receptor activation (Ibanez 1998). Neurotrophins play key roles in several biological processes in both the developing and adult nervous system. They also regulate development, maintenance, and function of peripheral (PNS) and central (CNS) nervous systems as well as modulate many aspects of neural function. In the mature nervous system, in addition to modulate neuronal survival, they also control synaptic function and synaptic plasticity. Studies have shown neurotrophins affect several aspects of synaptic transmission (Mizuno et al. 1994, Widmer and Hefti 1994, Tanaka et al. 1997, Rutherford et al. 1997). Structural changes in axonal and dendritic arbors are characteristic features of synaptic plasticity. Neurotrophins modulate these structural changes which lead to significant changes in the number and location of functional synapses (LeviMontalcini et al. 1954, Katz and Shatz 1996).

Synaptic plasticity represents one of the most fundamental and important functions of the brain (Holtmaat and Svoboda 2009, Kessels and Malinow, 2009, Yoshihara et al. 2009). This critical brain function has a key role in both short- and long-term memory. Neuroplasticity allows the neurons to compensate for injury as well as disease state and to adjust their activities in response to new situations or to change their environment. The mechanisms underlying these changes have been linked to the pathophysiology and treatment of multiple neurobiological disorders, including psychiatric disorder. Here we review the role of various neurotrophins in regulating synaptic plasticity and how the disruption of functional and structural connections of the neural circuits effects psychiatric disorder.

2. BDNF and Synaptic Plasticity in Psychiatric Disorder

BDNF play a crucial role many brain diseases. Studies have shown the involvement of BDNF in a wide range of neurophysiological processes. The characteristic of BDNF in the pattern of synthesis (which involves several biologically active isoforms that interact with different receptors) helps in controlling numerous signaling pathways. BDNF is present in nearly all brain regions (Hofer et al. 1990; Yan et al. 1997). The function of BDNF differs depending on both the stage of brain development as well as the constituents of the brain tissue (neuronal, glial, or vascular). The pivotal functions of BDNF is during developmental processes, regulation of neuro-, glio-, synaptogenesis, neuroprotection as well as regulation of short- and long-term synaptic plasticity that influence mechanisms of higher cognitive functions including learning and memory. BDNF has an important role in neurogenesis which is shown by the binding of BDNF to TrkB receptor. This binding stimulates neuronal differentiation and dendritic morphogenesis in the subgranular zone of the hippocampus, confirming its function during advance stages of neurogenesis. In one study, it has been explained that BDNF deficit does not result in significant decrease in the number of neurons, however it causes inhibition of dendritic arborization and deteriorates synaptic plasticity (Gao et al. 2009; Rauskolb et al. 2010; Wang et al. 2015). Further, decreased expression of genes, that are functionally related to vesicular trafficking and synaptic communication is observed with reduction of BDNF concentration induced in cultured rat hippocampal neurons (Mariga et al. 2015). Studies both in vivo and in vitro have shown the extensive role of BDNF in regulating synaptic plasticity and various learning mechanisms (Messaoudi et al. 2002; Minichiello et al. 1999). BDNF effects both functional and structural aspects of synaptic transmission by increasing the activity induced changes, which leads to enhanced efficiency of pathways influencing signal transduction (Lynch 2004). Reports have indicated that that learning increases BDNF gene expression, pro-BDNF and TrkB protein level in hippocampus (Silhol et al. 2007). Moreover, voluntary exercises correlated with improved performance in the Morris water maze test and other behavioral tasks related to learning alongwiththe increased level of BDNF in mouse hippocampus is observed (Vaynman et al. 2004). Similar observation is also reported by other researchers showing a correlation between physical activity-evoked BDNF overexpression and augmented excitatory transmission (Canossa et al. 1997; Castren et al.1993; Patterson et al. 1992; Zafra et al. 1990), as well as enhanced synaptic plasticity in the dentate gyrus (Lynch 2004). Further, prenatal sound stimulation increases the levels of BDNF (Chaudhury and Wadhwa 2009), synaptic proteins (Chaudhury et al. 2010) as well as number of synapses (Chaudhury et al. 2009) in chick hippocampus. Thus stimulation results in increased expression of BDNF which then strengthens synaptic potentiation, modulates the axo-dendritic morphology, and positively influences neuronal activity.

BDNF can modify molecular processes responsible for hippocampal synaptic potentiation which is categorized, as short and long term (Abraham 2003; Kandel 2001; Leal et al. 2015; Sweatt 1999). BDNF controls the short-lasting processes by regulation of neurotransmitter release, modification of pre-existing proteins or synapse structure (Leal et al. 2015). The long-term effects of BDNF are related to changes in gene expression or protein synthesis (Korte et al. 1998; Park and Poo 2013). BDNF influences synaptic plasticity at both pre- or postsynaptic sites. At presynaptic sites, BDNF releases neurotransmitters and at postsynaptic sites, it changes the expression of receptors, their molecular characteristics, as well as regulation of signaling pathways to regulate synaptic plasticity (Edelmann et al. 2014). Reports suggests that the presynaptic effect of BDNF on hippocampal LTP was observed in Schaffer's collaterals (Zakharenko et al. 2003). Apart from role of BDNF in influencing synaptic plasticity through molecular mechanisms involving changes in receptor expression, BDNF also enhances the activity-dependent efficiency of synaptic transmission by inducing structural modifications. The BDNF/TrkB receptor signaling cascade improves the efficiency of synaptic transmission by triggering processes which leads to increase in the number of dendritic spines and increased dendritic arborization(Amaral and Pozzo-Miller 2007; Gonzalez et al. 2016; Kumar et al. 2005). Further, BDNF in cooperation with neurotransmitters, helps in modulating synaptic plasticity. Studies suggest that there is a reciprocal and modulatory relationship between BDNF and NO, which effectively influences synaptic plasticity (Biojone et al. 2015). Thus, from all these studies we can summarize that BDNF has a pivotal role to play in modulating synaptic plasticity. However, as synaptic plasticity is one of the major function in psychiatric disorder, BDNF also plays an important role as an activity-dependent factor. Higher levels of BDNF immunoreactivity, in patients receiving antidepressant treatments compared to non-treated individuals, is observed in hippocampus of postmoterm brain(Chen et al. 2001). Also, in the hippocampus of suicidal subjects decreased mRNA and protein levels of

BDNF and TrkB receptors is observed (Dwivedi et al. 2003). This studies projects the potential role of BDNF in psychiatric disorder especially in depression. Furthermore, BDNF levels are also detectable in the blood, where it is mainly stored in platelets (Fujimura et al. 2002). Moreover, studies have shown that BDNF serum levels appear to reflect brain BDNF levels (Karege et al. 2002). Following these - and several other research, a decreased BDNF serum levels in depressed or suicidal individuals, as well as increased BDNF levels in antidepressant-medicated patients is reported (Sen et al. 2008; Brunoni et al. 2008). Reduced BDNF plasma levels alongwith the reduction in plasma tPA levels is observed in late-onset of geriatric depression (Shi et al. 2010). To add to these studies decreased serum levels of mBDNF in MDD patients are reported by three different laboratories (Yoshida et al. 2012; Zhou et al. 2013; Yoshimura et al. 2014). Therefore, BDNF which influences activity-dependent synaptic plasticity might be a critical factor in psychiatric illnesses.

3. Other Neurotrophins and Synaptic Plasticity in Psychiatric Disorder

Unlike BDNF role of NGF in modulating synaptic plasticity and its role in psychiatric disorder has not been extensively studied. NGF is also a neurotrophin which is a potential cognitive enhancer. Together with BDNF, NGF plays important roles in the maintenance of cholinergic neurons of the basal forebrain, which are critically involved in cognitive processes including synaptic plasticity (Mufson et al. 1999). Studies following NGF treatment in young rats does not reveal any improvement in spatial learning and memory, although positive effects are usually observed in cognitively impaired, aged, lesioned, or ethanol-treated rats (Frick et al. 1997; Frielingsdorf et al. 2006; Lukoyanov et al. 2003; Markowska et al. 1994, 1996; Pizzo and Thal 2004). Also, reports have shown that intrahippocampal injections of function-blocking NGF antibodies results in impair spatial memory (Conner et al. 2009) which suggests an involvement of this neurotrophin in memory formation. NT-3 and NT-4/5 are the neroptrophins which have been hardly studied in cognition. A role for the NT-3 is observed in olfactory discrimination learning and memory (Naimark et al. 2007) while the role of NT-4 appears essential for hippocampus and amygdala-dependent long-term contextual fear memory (Xie et al. 2000). More studies are required to understand the molecular mechanism of neurotrophins, other than BDNF, in regulating synaptic plasticity in different brain regions.

Role of NGF, NT-3 and NT-4/5 have been suggested in several psychiatric illnesses. There are reports which show evidence that chronic stress can lead to anxiety- and depression-like behaviors and may influence the distribution of NGF, both in animal models and in humans (Alleva and Aloe 1989, Connor and Dragunow 1998). In the adult brain, reduction of NGF and BDNF signaling may be involved in the pathophysiology of psychiatric disorders, such as depression (Cirulli and Alleva 2009, Castr´en 2005, Altar 1999, Castr´en et al. 2007, Duman and Monteggia 2006, Sen et al. 2008). Further, the role for NGF has been proposed in the etiopathogenesis of schizophrenia (Bersani et al. 2000). NGF can also modulate various psychostimulant. It has been shown that amphetamine administration may directly reduce NGF synthesis in selected rat brain areas including hippocampus, occipital cortex, and hypothalamus (Angelucci et al. 2007a). Conversely, the damage induced by amphetamine administration in dopaminergic areas of the rat brain is able to rescued by NGF (Chaturvedi et al. 2006). Further, cocaine-induced inhibition of PC12 proliferation (Tosk et al. 1996)have been reversed by NGF and cocaine reduces NGF serum levels (Angelucci et al. 2007b). In another study, cannabishas been shown to reduced NGF serum levels (Angelucci et al. 2008) and increased NGF serum levels in schizophrenic patients (Jockers-Scheru¨bl et al. 2003). There are reports that following heroin and opioids treatment, disruption of normal development of striatal cholinergic neurons and reduction of striatal NGF in rats occurs (Wu et al. 2001). Dysfunctions of NGF, BDNF, and NT-3 may contribute to impaired brain development, neuroplasticity, and synaptic ‘‘dysconnectivity’’ leading to the schizophrenic syndrome which is reported in preclinical and clinical data (Angelucci et al. 2004, 2005; Shoval and Weizman 2005).

SUMMARY

Therefore, it is clear that neurotrophins are important in modulating neural circuits and via altering the synaptic plasticity the neurotrophins play a key role in psychiatric disorder. This is supported by several facts. It is suggested that neurotrophins may be altered by administration of psychostimulants. With the role of neurotrophins in neuronal survival and synaptic plasticity it is proposed that one of the mechanisms of psychostimulant-induced neurotoxicity may be linked to alterations of neurotrophins. Moreover, altered neurotrophins may be a part of the pathogenesis of psychiatric disorders. Since, psychiatric disorders are common in drug users, one hypothesis could be that the psychostimulants can cause psychosis through interfering with neurotrophins synthesis in CNS neurons. The implication of this hypothesis is to prevent psychostimulant neurotoxicity that has the potential to be useful in the treatment of several psychiatric disorders. Also disruption in synaptic plasticity occurs in various psychiatric illnesses. Further, activity results in modification in synaptic plasticity by regulating neurotrophins. Thus there is a possibility that activity may enhance neurotrophin which can help in reorganizing the disruption in synaptic plasticity occurs in psychiatric disorder. Further in addition to BDNF, studies are required to evaluate other neurotrophins and its correlation with synaptic plasticity following drug treatment in psychiatric disorder.

REFERENCES

Abraham, W. C. (2003) How long will long-term potentiation last? *Philos. Trans. R Soc. Lond. B Biol. Sci.,* 358:735-744.

Alleva, E., Aloe, L. (1989) Physiological roles of NGF in adult rodents: a biobehavioral perspective. *Int. J. Comp. Psychol.*, 2: 213-230.

Altar, C. A. (1999) Neurotrophins and depression. *Trends Pharmacol. Sci.*, 20: 59-61.

Amaral, M. D., Pozzo-Miller, L. (2007) TRPC3 channels are necessary for brain derived neurotrophic factor to activate a nonselective cationic current and to induce dendritic spine formation. *J. Neurosci.,* 27: 5179-5189.

Angelucci, F., Brene`, S., Mathe`, A. A. (2005) BDNF in schizophrenia, depression and BDNF in schizophrenia, depression and corresponding animal models. *Mol. Psychiatry*, 10: 345-352.

Angelucci, F., Gruber, S. H., El Khoury, A., Tonali, P. A., Mathe´, A. A. (2007a) Chronic amphetamine treatment reduces NGF and BDNF in the rat brain. *Eur. Neuropsychopharmacol.*, 17, 756-762.

Angelucci, F., Mathe´, A. A., Aloe, L. (2004) Neurotrophic factors and CNS disorders: Findings in rodent models of depression and schizophrenia. *Prog. Brain Res.*, 146: 151-165.

Angelucci, F., Ricci, V., Pomponi, M., Conte, G., Mathe´, A. A., Tonali, P. A., Bria, P. (2007b) Chronic heroin and cocaine abuse is associated with decreased serum concentrations of the nerve growth factor and brain-derived neurotrophic factor. *J. Psychopharmacol.,* 21: 820-825.

Angelucci, F., Ricci, V., Spalletta, G., Pomponi, M., Tonioni, F., Caltagirone, C., Bria, P. (2008) Reduced serum concentrations of nerve growth factor, but not brain-derived neurotrophic factor, in chronic cannabis abusers. *Eur. Neuropsychopharmacol.*, 18: 882-887.

Bersani, G., Iannitelli, A., Fiore, M., Angelucci, F. and Aloe, L., "Data and hypotheses on the role of nerve growth factor and other neurotrophins in psychiatric disorders", *Medical Hypotheses*, vol. 55, no. 3, pp. 199-207, 2000.

Biojone, C., Casarotto, P. C., Joca, S. R., Castre´n, E. (2015) Interplay between nitric oxide and brain-derived neurotrophic factor in neuronal plasticity.*CNS Neurol. Disord.*, 14:979-987.

Brunoni, A. R., Lopes, M., Fregni, F. (2008) A systematic review and meta-analysis of clinical studies on major depression and BDNF levels:

implications for the role of neuroplasticity in depression. *Int. J. Neuropsychopharmacol.,* 11: 1169-1180.

Canossa, M., Griesbeck, O., Berninger, B., Campana, G., Kolbeck, R., Thoenen, H. (1997) Neurotrophin release by neurotrophins: implications for activity-dependent neuronal plasticity. *Proc. Natl. Acad. Sci. USA,* 94:13279-13286.

Castr´en, E., V˜oikar, V., Rantamäki, T. (2007) Role of neurotrophic factors in depression", *Curr. Opin. Pharmacol.,* 7: 18-21.

Castren, E., Pitkanen, M., Sirvio, J., Parsadanian, A., Lindholm, D., Thoenen, H., Riekkinen, P. J. (1993) The induction of LTP increases BDNF and NGF mRNA but decreases NT-3 mRNA in the dentate gyrus. *Neuroreport,* 4:895-898.

Chaturvedi, R. K., Shukla, S., Seth, K., Agrawal, A. K. (2006) Nerve growth factor increasessurvival of dopaminergic graft, rescue nigral dopaminergic neurons and restores functional deficits in rat model of Parkinson's disease. *Neurosci. Lett.*, 39844-39849.

Chaudhury, S., Jain, S., Wadhwa, S. (2010) Expression of synaptic proteins in the hippocampus and spatial learning in chicks following prenatal auditory stimulation. *Dev. Neurosci.*, 32: 114-124.

Chaudhury, S., Nag, T. C., Wadhwa, S. (2009) Effect of prenatal auditory stimulation on numerical synaptic density and mean synaptic height in the posthatch Day 1 chick hippocampus. *Synapse,* 63:152-159.

Chaudhury, S., Wadhwa, S. (2009) Prenatal auditory stimulation alters the levels of CREB mRNA, p-CREB and BDNF expression in chick hippocampus. *Int. J. Dev. Neurosci.*, 27: 583-590.

Chen, B., Dowlatshahi, D., MacQueen, G. M., Wang, J. F., Young, L. T. (2001) Increased hippocampal BDNF immunoreactivity in subjects treated with antidepressant medication. *Biol. Psychiatry*, 50: 260-265.

Chen, B., Dowlatshahi, D., MacQueen, G. M., Wang, J. F., Young, L. T. (2001) Increased hippocampal BDNF immunoreactivity in subjects treated with antidepressant medication. *Biol. Psychiatry*, 50: 260-265.

Cirulli, F., Alleva, E. (2009) The NGF saga: from animal models of psychosocial stress to stress-related psychopathology. *Front Neuroendocrinol.*, 30: 379-395.

Connor, B., Dragunow, M. (1998) The role of neuronal growth factors in neurodegenerative disorders of the human brain. *Brain Res. Rev.*, 27: 1-39.

Duman, R. S., Monteggia, L. M. (2006) A neurotrophic model for stress-related mood disorders. *Biol. Psychiatry*, 59: 1116-1127.

Dwivedi, Y., Rizavi, H. S., Conley, R. R., Roberts, R. C., Tamminga, C. A., Pandey, G. N. (2003) Altered gene expression of brain-derived neurotrophic factor and receptor tyrosine kinase B in postmortem brain of suicide subjects. *Arch. Gen. Psychiatry*, 60: 804-815.

Edelmann, E., Lessmann, V., Brigadski, T. (2014) Pre- and postsynaptic twists in BDNF secretion and action in synaptic plasticity. *Neuropharmacology,* 76: 610-627.

Frick, K. M., Price, D. L., Koliatsos, V. E., Markowska, A. L. (1997) The effects of nerve growth factor on spatial recent memory in aged rats persist after discontinuation of treatment. *J. Neurosci.*, 17: 2543-2550.

Frielingsdorf, H., Thal, L. J., Pizzo, D. P. (2006) The septohippocampal cholinergic system and spatial working memory in the Morris water maze. *Behav. Brain Res.,* 168: 37-46.

Fujimura, H., Altar, C. A., Chen, R., Nakamura, T., Nakahashi, T., Kambayashi, J., Sun, B., Tandon, N. N. (2002) Brain-derived neurotrophic factor is stored in human platelets and released by agonist stimulation. *Thromb. Haemost.*, 87: 728-734.

Gao, X., Smith, G. M., Chen, J. (2009) Impaired dendritic development and synaptic formation of postnatal-born dentate gyrus granular neurons in the absence of brain-derived neurotrophic factor signaling. *Exp. Neurol.*, 215: 178-190.

Gonzalez, A., Moya-Alvarado, G., Gonzalez-Billaut, C., Bronfman, F. C. (2016) Cellular and molecular mechanisms regulating neuronal growth by brain-derived neurotrophic factor (BDNF). *Cytoskeleton (Hoboken),* 73: 612-628.

Hallbook, F. (1999) Evolution of the vertebrate neurotrophin and Trk receptor gene families. *Curr. Opin. Neurobiol.*, 9: 616-621.

Hellweg, R. (2003) Chronic cannabis abuse raises nerve growth factor serum concentrations in drug-naive schizophrenic patients. *J. Psychopharmacol.*, 17: 439-445.

Hofer, M., Pagliusi, S. R., Hohn, A., Leibrock, J., Barde, Y. A. (1990) Regional distribution of brain-derived neurotrophic factor mRNA in the adult mouse brain. *EMBO J.,* 9: 2459-2464.

Holtmaat, A., Svoboda, K. (2009) Experience-dependent structural synaptic plasticity in the mammalian brain. *Nat. Rev. Neurosci.,* 10: 647-658.

Ibanez, C. F. (1998) Emerging themes in structural biology of neurotrophic factors. *Trends Neurosci.,* 21: 438-444.

Jockers-Scherübl, M. C., Matthies, U., Danker-Hopfe, H., Lang, U. E., Mahlberg, R., Kandel, E. R. (2001) Themolecular biology ofmemory storage: a dialogue between genes and synapses. *Science,* 294:1030-1038.

Karege, F., Schwald, M., Cisse, M. (2002) Postnatal developmental profile of brain-derived neurotrophic factor in rat brain and platelets. *Neurosci. Lett.,* 328: 261-264.

Katz, L. C., Shatz, C. J. (1996) Synaptic activity and the construction of cortical circuits. *Science,* 274: 1133-38.

Kessels, H. W., Malinow, R. (2009) Synaptic AMPA receptor plasticity and behavior. *Neuron,* 61: 340-350.

Korte, M., Kang, H., Bonhoeffer, T., Schuman, E. (1998) A role for BDNF in the late-phase of hippocampal long-term potentiation. *Neuropharmacol.*, 37: 553-559.

Kumar, V., Zhang, M. X., Swank, M. W., Kunz, J., Wu, G. Y. (2005) Regulation of dendritic morphogenesis by Ras-PI3K-Akt-mTOR and Ras-MAPK signaling pathways. *J. Neurosci.,* 25: 11288-11299.

Leal, G., Afonso, P. M., Salazar, I. L., Duarte, C. B. (2015) Regulation of hippocampal synaptic plasticity by BDNF. *Brain Res.,* 1621: 82-101.

Levi-Montalcini, R., Meyer, H., Hamburger, V. (1954) In vitro experiments on the effects of mouse sarcomas 180 and 37 on the spinal and sympathetic ganglia of the chick embryo. *Cancer Res.,* 14: 49-57.

Lukoyanov, N. V., Pereira, P. A., Paula-Barbosa, M. M., Cadete-Leite, A. (2003) Nerve growth factor improves spatial learning and restores hippocampal cholinergic fibers in rats withdrawn from chronic treatment with ethanol. *Exp. Brain Res.*, 148: 88-94.

Lynch, M. A. (2004) Long-term potentiation and memory. *Physiol. Rev.,* 84:87-136.

Mariga, A., Zavadil, J., Ginsberg, S. D., Chao, M. V. (2015) Withdrawal of BDNF from hippocampal cultures leads to changes in genes involved in synaptic function. *Dev. Neurobiol.*, 75: 173-192.

Markowska, A. L., Koliatsos, V. E., Breckler, S. J., Price, D. L., Olton, D. S. (1994) Human nerve growth factor improves spatial memory in aged but not in young rats. *J. Neurosci.,* 14: 4815-4824.

Messaoudi, E., Ying, S. W., Kanhema, T., Croll, S. D., Bramham, C. R. (2002) Brain-derived neurotrophic factor triggers transcription-dependent, late phase long-term potentiation in vivo. *J. Neurosci.*, 22: 7453-7461.

Minichiello, L., Korte, M., Wolfer, D., Ku¨hn, R., Unsicker, K., Cestari, V., Rossi-Arnaud, C., Lipp, H. P., Bonhoeffer, T., Klein, R. (1999) Essential role for TrkB receptors in hippocampus-mediated learning. *Neuron,* 24:401-414.

Naimark, A., Barkai, E., Matar, M. A., Kaplan, Z., Kozlovsky, N., Cohen, H. (2007) Upregulation of neurotrophic factors selectively in frontal cortex in response to olfactory discrimination learning. *Neural Plast.,* 2007: 13427.

Mufson, E. J., Kroin, J. S., Sendera, T. J., Sobreviela, T. (1999) Distribution and retrograde transport of trophic factors in the central nervous system: Functional implications for the treatment of neurodegenerative diseases. *Prog. Neurobiol.,* 57: 451-484.

Park, H., Poo, M. M. (2013) Neurotrophin regulation of neural circuit development and function. *Nat. Rev. Neurosci.*, 14: 7-23.

Patterson, S. L., Grover, L. M., Schwartzkroin, P. A., Bothwell, M. (1992) Neurotrophin expression in rat hippocampal slices: a stimulus paradigm inducing LTP in CA1 evokes increases in BDNF and NT-3 mRNAs. *Neuron,* 9: 1081-1088.

Pizzo, D. P., Thal, L. J. (2004) Intraparenchymal nerve growth factor improves behavioral deficits while minimizing the adverse effects of intracerebroventricular delivery. *Neuroscience*, 124: 743-755.

Rauskolb, S., Zagrebelsky, M., Dreznjak, A., Deogracias, R., Matsumoto, T., Wiese, S., Erne, B., Sendtner, M., Schaeren-Wiemers, N., Korte, M., Barde, Y. A. (2010) Global deprivation of brain-derived neurotrophic factor in the CNS reveals an area-specific requirement for dendritic growth. *J. Neurosci.,* 30: 1739-1749.

Sen, S., Duman, R., Sanacora, G. (2008) Serum brain-derived neurotrophic factor, depression, and antidepressant medications: meta-analyses and implications. *Biol. Psychiatry*, 64: 527-532.

Shi, Y., You, J., Yuan, Y., Zhang, X., Li, H., Hou, G. (2010) Plasma BDNF and tPA are associated with late-onset geriatric depression. *Psychiatry Clin. Neurosci*., 64: 249-254.

Shoval, G., Weizman, A. (2005)The possible role of neurotrophins in the pathogenesis and therapy of schizophrenia. *Eur. Neuropsychopharmacol.,* 15: 319-329.

Silhol, M., Arancibia, S., Maurice, T., Tapia-Arancibia, L. (2007) Spatial memory training modifies the expression of brain-derived neurotrophic factor tyrosine kinase receptors in young and aged rats. *Neuroscience,* 146: 962-973.

Sweatt, J. D. (1999) Toward a molecular explanation for long-term potentiation. *Learn Mem*., 6: 399-416.

Vaynman, S., Ying, Z., Gomez-Pinilla, F. (2004) Hippocampal BDNF mediates the efficacy of exercise on synaptic plasticity and cognition. *Eur. J. Neurosci.,* 20: 2580-2590.

Wang, L., Chang, X., She, L., Xu, D., Huang, W., Poo, M. M. (2015) Autocrine action of BDNF on dendrite development of adult-born hippocampal neurons. *J. Neurosci.,* 35: 8384-8393.

Wu, V. W., Mo, Q., Yabe, T., Schwartz, J. P., Robinson, S. E. (2001) Perinatal opioids reduce striatal nerve growth factor content in rat striatum. *Eur. J. Pharmacol*., 41: 211-214.

Xie, C. W., Sayah, D., Chen, Q. S., Wei, W. Z., Smith, D., Liu, X. (2000) Deficient long-term memory and long-lasting long-term potentiation in mice with a targeted deletion of neurotrophin-4 gene. *Proc. Natl. Acad. Sci. USA,* 97: 8116-8121.

Yan, Q., Rosenfeld, R. D., Matheson, C. R., Hawkins, N., Lopez, O. T., Bennett, L., Welcher, A. A. (1997) Expression of brain-derived neurotrophic factor protein in the adult rat central nervous system. *Neuroscience,* 78: 431-448.

Yoshida, T., Ishikawa, M., Niitsu, T., Nakazato, M., Watanabe, H., Shiraishi, T., Shiina, A., Hashimoto, T., Kanahara, N., Hasegawa, T., Enohara, M., Kimura, A., Iyo, M., Hashimoto, K. (2012) Decreased serum levels of mature brain-derived neurotrophic factor (BDNF), but not its precursor proBDNF, in patients with major depressive disorder. *PLoS ONE*, 7: e42676.

Yoshihara, Y., De Roo, M., Muller, D. (2009) Dendritic spine formation and stabilization. *Curr. Opin. Neurobiol*., 19: 146-153.

Yoshimura, R., Kishi, T., Hori, H., Atake, K., Katsuki, A., Nakano-Umene, W., Ikenouchi-Sugita, A., Iwata, N., Nakamura, J. (2014) Serum proBDNF/BDNF and response to fluvoxamine in drug-na€ıve firstepisode major depressive disorder patients. *Ann. Gen. Psychiatry,* 13: 19.

Zafra, F., Hengerer, B., Leibrock, J., Thoenen, H., Lindholm, D. (1990) Activity dependent regulation of BDNF and NGF mRNAs in the rat hippocampus is mediated by non-NMDA glutamate receptors. *EMBO J.,* 9: 3545-3550.

Zakharenko, S. S., Patterson, S. L., Dragatsis, I., Zeitlin, S. O., Siegelbaum, S. A., Kandel, E. R., Morozov, A. (2003) Presynaptic BDNF required for a presynaptic but not postsynaptic component of LTP at hippocampal CA1-CA3 synapses. *Neuron,* 39: 975-990.

Zhou, L., Xiong, J., Lim, Y., Ruan, Y., Huang, C., Zhu, Y., Zhong, J. H., Xiao, Z., Zhou, X. F. (2013) Upregulation of blood proBDNF and its receptors in major depression. *J. Affect Disord.,* 150: 776-784.

In: Synaptic Plasticity
Editor: Sraboni Chaudhury

ISBN: 978-1-53613-606-7

Chapter 8

HIPPOCAMPAL SYNAPTIC PLASTICITY: INSIGHT OF EPILEPSY PROGRESSION, DEVELOPMENT AND PUTATIVE TARGETS FOR TREATMENT

Manisha Shrivastav[1,*] and Dilip Verma[2]
[1]Epilepsy Research Center, Westfälische Wilhelms-Universität Münster, Germany
[2]Institute of Physiology I, Westfälische Wilhelms-Universität Münster, Germany

ABSTRACT

Synapses are the basis of neuronal functions and synaptic plasticity is the ability to strengthen or weaken the synapses. Therefore, directly or indirectly synaptic plasticity plays crucial role in many neuropathophysiological conditions though subcellular alterations varying for each disorder. In last decade, it has been shown that neuronal

* Corresponding Author Email: manisha.shrivastava@gmail.com.

networks are homeostatically regulated, dynamic and show time dependent alterations, which changes neuronal activity. It has been also shown that synaptic homeostasis and plasticity could not only provide insight of several neurological disorders like migraine, Alzheimer's disease, epilepsy but could also provide better treatment and management strategies. Resent reports suggested that epileptic stages could overlay with the mechanisms to alter the factors, contributing to the synaptic plasticity of excitatory and inhibitory circuits. In this chapter, we will focus on synaptic plasticity and homeostasis of synaptic plasticity in epilepsy progression and putative targets for treatment.

Keyword: synapse, synaptic plasticity, epilepsy

1. Introduction

Synapse is a structure located at the junction of a neuron to transmit the signals to other neuron. Whereas, synaptic plasticity is the synapse competence to either decrease or increase the response to further signal transmission. Previously, synaptic plasticity used to describe learning and memory. However, recent research uses the synaptic plasticity concept as a tool to define the normal physiology or pathological conditions. Synapses are crucial for normal neuronal communication. Therefore, it might be right to say that alterations in synapse composition might contribute to pathophysiology. Perhaps, synaptic modulation leads to different mode of communication or to alter signal intensity by increasing the inhibition or decreasing the inhibition. Furthermore, alteration in synaptic plasticity could be multifactorial, which could change the output of communication to develop pathological conditions like epilepsy, migraine with aura etc.

Epilepsy is a condition, associated with irregular electrical activity of neurons in the brain. Subsequently, abnormal electrical activities generate recurrent episodes of sensory disturbance, loss of consciousness or convulsions. Previous research has shown connection between synaptic plasticity homeostasis and epileptogenesis. To understand the connection between the two, it is important to get insight view of neuronal homeostasis. Therefore, in this chapter first, we will describe the homeostasis and later we will focus on how synaptic plasticity and synaptic homeostasis may contribute to epilepsy.

2. HOMEOSTASIS IN NEURONS

Homeostasis is important for the nervous system to maintain and sometimes even to restore the normal physiological activity in the presence of perturbation. Homeostasis is a process that establishes the equilibrium for normal physiological functions. Homeostasis is required for the maintenance of neuronal activity in physiological range. Whether homeostasis is related to the neurological disorders? If it is related, it is useful to explore the homeostasis contributing factors in neurons and their participation in homeostasis. How homeostasis could contribute to the alteration of neuronal plasticity? Here, we will discuss all of these questions in a precise manner.

The homeostasis model of neuronal system proposes, cellular monitors, which observe the changes in activity and innervation. Whereas, intracellular and intercellular signaling pathways transducing the information to fine-tune synaptic and ion channel's functions. Previous reports have suggested that neurons are capable to adapt according to the changes in synaptic alterations and synaptic function within physiological range (Davis and Goodman, 1998a). There are many mechanisms of homeostasis regulation in nervous system. These regulations are associated to presynaptic, postsynaptic, autonomous, intercellular or intracellular signaling induced alterations in ion channel density and neurotransmitter receptor expression (Bliss and Collingridge, 1993; Katz and Shatz, 1996; Davis and Goodman, 1998b; McAllister et al., 1999; Sanes and Lichtman, 1999; Turrigiano, 1999). Homeostatic regulation compensates the effect of postsynaptic excitability, disease or injury and altered innervation by three ways: 1) controlling the intrinsic excitability, 2) expression of neurotransmitter receptors, 3) control of neurotransmitter release (O'Brien et al., 1998; Turrigiano et al., 1998; Turrigiano and Nelson, 2004; Davis, 2006; Frank et al., 2006; Nerbonne et al., 2008; Marder, 2011; Temporal et al., 2011; Turrigiano, 2011; Wenner, 2011; Davis, 2013; Parrish et al., 2014). Homeostatic signaling can regulate the presynaptic transmitter release from 10-200% (Frank et al., 2006), which can be sustained from seconds to days or even years (Plomp et al., 1992). This adjustment is synapse specific. There are some molecules monitoring cellular activity, excitability and plasticity.

2.1. Cellular Monitors to Establish Homeostasis

There are molecules which act like cell monitoring system, noticing cellular events precisely (activity, excitability and plasticity changes) to create homeostasis (Davis and Bezprozvanny, 2001). Calcium enters in neurons through voltage gated channels, neurotransmitter receptors and also released from internal stores, which is crucial for vesicle release, cell signaling and gene transcription. This calcium entry is also generating calcium waves and oscillations which gives temporal information about changes in cell activity and is also source of spatial information regarding subcellular localization and diffusion within the cell (De Koninck, 1998; Crabtree, 1999; Graef et al., 1999).

2.1.1. CaMKII

Ca^{2+}/ calmodulin-dependent protein kinase II (CaMKII) is a serine/ threonine specific protein kinase, which is regulated by Ca^{2+}/ calmodulin complex. It monitors calcium spikes frequency (Schulman et al., 1995; De Koninck, 1998) and also necessary for Ca^{2+} homeostasis. It is playing important role to modulate neuronal excitability, synaptic function and gene expression (Srinivasan et al., 1994; Thomas et al., 1994; Braun and Schulman, 1995; Deisseroth et al., 1998; Rongo and Kaplan, 2000). CaMKII activity is dependent on CaM concentration, phosphatase activity, enzyme localization and its subunit composition (Braun and Schulman, 1995; Brocke et al., 1995; Schulman et al., 1995). Any of these variables are sufficient to alter CaMKII activity. In neurons CaMKII consist 6-12 alpha and beta subunits, either in homomeric or in heteromeric configuration. Function of CaMKII is modulated by subunit composition (Brocke et al., 1995; De Koninck, 1998). Altered subunit composition can also regulate the CaMKII subcellular targeting and its sensitivity (Srinivasan et al., 1994).

2.1.2. Calcineurin

Calcineurin is a calcium and calmodulin dependent serine/ threonine protein phosphatase. It can cause calcium oscillation specific changes to gene transcription (Meldolesi, 1998; Crabtree, 1999; Graef et al., 1999). It can change transcriptional selectivity and magnitude of transcriptional response by sensing the calcium oscillation specific changes in the neuron. Calcineurin plays a role in inhibitory synaptic functions. Thus, altering the neuronal excitability after LTP (Long term potentiation) (Mulkey et al., 1994; Lai et al., 1999; Wan, 1999; Lu et al., 2000).

2.1.3. Calcium Channels

Voltage gated calcium channels also act like calcium sensor and participating in neuronal homeostasis. L-type calcium channels offer calcium entry in neurons to trigger the calmodulin translocation to the nucleus and activation of cAMP response element-binding protein-mediated transcription (CREB) (Bito et al., 1996; Deisseroth et al., 1996; 1998). L-type calcium channels activated by excitatory postsynaptic potential (EPSP) stimulation due to their slow activation kinetics. L-type calcium channel requires relatively high negative potential to activate, which minimize the possibility of their contribution in action potential (Mermelstein et al., 2000). EPSPs define the location of depolarization with the dendritic tree, whereas, action potential is the summed cell depolarization. Calcium entry can transduce the calcium mediated signaling, facilitating the decoding of neuronal voltage activity into chemical signaling.

2.2. Presynaptic Homeostasis Regulation

There are two types of neuronal homeostatic signaling: 1) intrinsic plasticity, plasticity within the neurons and 2) synaptic scaling, plasticity between the neurons (Davis and Bezprozvanny, 2001; Eichler and Meier, 2008; Turrigiano, 2011). Intrinsic plasticity is a continuous change in neuronal intrinsic electrical property by synaptic plasticity (Desai et al., 1999; Karmarkar and Buonomano, 2006). It is a component of homeostasis in neurons that regulates the electrical activity. Expression level of ion channels and changes in ion channels biophysical properties are contributing to the intrinsic plasticity, which is affecting further electrical properties of neurons e.g., spike generation, propagation, backward propagation, as well as, metaplasticity (Abraham and Bear, 1996). Synaptic scaling is homeostatic plasticity, regulating the firing rate of action potential in a neuron.

2.2.1. Ion Channel

Ion channels play crucial role not only to define the electrical activity of neurons but also in homeostasis. Neurons are fine tuning the excitability by shifting the ion channel's location (Grubb and Burrone, 2010). It has been observed that voltage gated N^+ channels shift their location during chronic depolarization (Grubb and Burrone, 2010). This is a reversible process, which is also dependent on the activity of voltage gated T/ L-type Ca^{2+} channels. Alternative splicing regulates polarized targeting of $K_v3.1$ channels also regulating the neuronal firing rates (Jonas, 2000; Gu et al., 2011). Previous research showed that $K_v1.1$ and 1.2 are also crucial to regulate neuronal excitability (Wang et al., 1994). K_v channels participate in feedforward inhibitory circuits, involved in suppression of excitability and also regulating the temporal information for action potential (Lawrence and McBain, 2003). In temporal lobe epilepsy (TLE) mouse model, upregulation of $K_v1.1$ transcription factor in granule cells has been determined (Kirchheim et al., 2013). Furthermore, both mouse models $K_v1.1$ knockout (Wenzel et al., 2007) and $K_v1.1$ lentiviral overexpressed mouse develops epilepsy. $K_v1.1$ mouse shows abnormal interictal activities in electroencephalography (EEG) and video recording. Whereas, lentiviral overexpression of $K_v1.1$ reduces the intrinsic excitability of transduced hippocampal pyramidal neurons (Wykes et al., 2012). Therefore, ion channels are very important components of neuronal homeostasis.

2.2.2. Glycine Receptor

Glycine receptors (GlyR) spontaneous activity increasing the presynaptic excitability and functions of Glutamatergic neurons. There are several reports defining the impact of glycine receptor mutation (in ligand binding domain), which leads gain of function and contribute to spontaneous channel activity (Meier et al., 2005; Eichler et al., 2009; Legendre et al., 2009; Kletke et al., 2013; Winkelmann et al., 2014). Beside that, spontaneous activity of GlyR can also alter neurotransmitter release to contribute in synaptic homeostasis (Zeilhofer et al., 2004).

2.2.3. Retrograde Autocrine Signaling

In both excitatory and inhibitory neurons, cannabinoid receptors (CB) are expressed presynaptically (Alger, 2005; Katona and Freund, 2008; Hill et al., 2011) and show different impact on neuronal excitability. In inhibitory neurons, CB1 receptors are activated by retrograde autocrine signaling (post to presynaptic release of CBs anandamide or 2-AG) and decrease the gamma aminobutyric acid (GABA) release. Moreover, it has been observed that K^+ inward rectifier (Kir) induced K^+ current increased via CB1 activation (Mackie et al., 1995; Chemin et al., 2001). Whereas, in excitatory neurons, activation of CB1 leads reduced glutamate release (Domenici et al., 2006; Kawamura et al., 2006). This could explain the possible role of CB1 in neuronal homeostasis (Ohno-Shosaku et al., 2001; Wilson et al., 2001; Chen et al., 2003; Blair et al., 2006; Chen et al., 2007; Bojnik et al., 2011; Goffin et al., 2011; Karlócai et al., 2011).

2.2.4. Adenosine Paracrine Signaling

Astrocytes are releasing adenosine triphosphate (ATP), which can either directly enhance neuronal excitability via purinergic receptors or can reduce glutamate release, indirectly via signaling through adenosine A1 receptors (Nicoll and Schmitz, 2005; Boison, 2013; Dias et al., 2013). Consequently, adenosine signaling is one of the crucial components of presynaptic homeostasis, which has significance in epileptogenesis (Boison, 2011).

2.2.5. Homeostatic Control of Neurotransmitter Release

Postsynaptic modulation of neurotransmitter receptor expression level affecting the quantal release of neurotransmitter, in both excitatory and inhibitory neurons (Turrigiano, 2011; Wenner, 2011). Several studies showed that in homeostasis, neurons have modified synaptic ultrastructure, action potential (AP)-driven presynaptic Ca^{2+} influx and vesicle recycling (Murthy et al., 2001; Kim and Ryan, 2010; Zhao et al., 2011).

2.3. Post Synaptic Homeostasis Regulation

Research showed that increased presynaptic release can restore normal depolarization in neuromuscular junction regardless of impaired postsynaptic receptors. Postsynaptic receptor function of neuromuscular junction can be compensated by increased presynaptic release, which leads to normal muscle depolarization (Davis et al., 1998; Marek et al., 2000).

2.3.1. Homeostasis in Postsynaptic Excitability

Central nervous system neurons regulate synaptic activity by modulating the postsynaptic neurotransmitter receptor expression in response to the activity level. Experimental data suggested that postsynaptic sensitivity is altered according to the excitatory or inhibitory input (O'Brien et al., 1998; Turrigiano et al., 1998).

2.3.2. Homeostasis in Altered Innervation

Homeostatic regulation maintains the overall synaptic input within a physiological range by altering the synaptic plasticity, which is also known as synaptic scaling (Liu and Tsien, 1995; Davis and Goodman, 1998a; 1998b). Thus, synaptic strengthening or weakening is dependent on total sum of physiological range of activity. For example, if number of synapses are increased then to maintain the activity in normal physiological range, the synaptic efficacy will be reduced and vice-versa.

2.3.3. Control of Postsynaptic Neurotransmitter Receptors

Neurons modulate the expression of neurotransmitter receptors to maintain the normal physiological activity (Turrigiano, 2000). Postsynaptic response alters in response to the excitatory or inhibitory input to the neuron. It is an activity driven phenomenon, regulated by the postsynaptic neurotransmitter receptor expression level (Carroll et al., 1999; Lüscher et al., 1999; Shi, 1999; Hayashi, 2000).

2.3.4. Postsynaptic Sensors, Scaffolds and Signaling

As previous studies have shown, presynaptic neurotransmitter release alteration is dependent to postsynaptic neurotransmitter receptor disruption (Davis, 2013). Therefore, system is required, which will detect the changes and transmit the information to presynaptic neuron. Henry et al. showed that mTORC1 signaling is contributing as a retrograde signaling system to regulate the presynaptic release to support the homeostasis (Henry et al., 2012). Drosophila neuromuscular junctions (NMJ) research studies provide information regarding the sensors, scaffolding proteins and signaling pathways that contributes to the homeostasis of neuronal functions. These factors may function in mammalian central nervous system too (Frank et al., 2009; Giagtzoglou et al., 2009; Pilgram et al., 2011; Henry et al., 2012; Laplante and Sabatini, 2012; Bateup et al., 2013; Frank, 2013).

2.4. Homeostasis in Disease or Injury

Homeostasis hypothesis suggests, disturbance in homeostasis process or failure to maintain the neuronal activity may contribute to the neurological disorders. Interruption in homeostatic regulation might lead functional alterations of neuronal circuits that cause seizure-like activity or neuronal insult. As an example, axonal lesions in hippocampal slices induces alterations in axonal sprouting and hyperexcitability cause posttraumatic epilepsy (McKinney et al., 1997).

2.5. Homeostatic Synaptic Regulation

A mechanism is required to monitor total summed excitation in neurons to create homeostasis. One possible mechanism is Calcium-dependent signaling, which may contribute to sense action potential induced depolarization of neurons. Therefore, soma and other central locations in neurons have monitoring molecules to detect action potential induced activity. Second mechanism could be that homeostasis mechanism is a retrograde signaling where, monitors are localized at postsynaptic synapses and may be relaying the information via back propagation of action potential (Bi and Poo, 1998).

2.6. Trans-Synaptic Signal

Studies suggested that neurotrophins and nerve growth factors are serving as trans-synaptic signaling molecules, which contribute to the synaptic plasticity (McAllister et al., 1999). Whereas, biochemical experimental data showed that proteins localized at pre and postsynaptic densities are playing crucial role in trans-synaptic signaling (Butz et al., 1998; Maximov et al., 1999). Neuraxin localized at presynaptic densities have single transmembrane domain and a short carboxy end with protein 4.1 and CASK-PDZ domain-binding sequence motifs (Hata et al., 1996). While, neuroligin is present at postsynaptic densities, which has cytoplasmic domain with a PDZ-binding motif with ability to bind synaptic adapter proteins as PSD-95 (Irie et al., 1997). Thus, these proteins are binding to modular adapter proteins CASK and PSD-95 via PDZ-mediated interactions. Neuraxin-neuroligin asymmetric junction may play role in synapse organization to contribute to trans-synaptic signaling via trans-synaptic scaffold protein around them. This may be regulating expression or function of presynaptic calcium channels. The core element of trans-synaptic protein complex can modify transcription. Trans-synaptic singling contributes to synapse growth, their stabilization and activity dependent plasticity (Martin and Kandel, 1996; Schuster et al., 1996; Davis et al., 1997; Bredt, 2000; Tanaka et al., 2000; Broadie, n.d.).

3. EPILEPSY

Epilepsy considered as a chronic neurological disorder characterized by recurrent seizures. According to the World Health Organization 2012, in general 4-10 per 1000 people are affected with active epilepsy (Kandratavicius et al., 2014). Categorically, there are two types of seizures, 1) generalized seizures exhibited by widespread electrical discharge in both cerebral hemispheres, typically linked with genetic factors and 2) partial seizure exhibited by a local electrical discharge in the brain, typically caused by local brain injuries e.g., stroke and tumor.

There are several human epilepsies all together grouped under the definition of epileptic encephalopathies (EEs) (Sharma et al., 2007). This book chapter is mainly focused on TLE, which is the most common form of epilepsy in humans. This form of epilepsy includes focal seizures and the seizures spread to neighboring cortices. The pathophysiology of TLE is vary according to the location of the abnormality. Mesial temporal or limbic epilepsy (MLTE) is also known as Limbic epilepsy or hippocampal sclerosis. It is one of the most common forms of focal epilepsy and most frequently operated in the clinics (Sharma et al., 2007). Unlike other types of epilepsy, which vary in pathology and location, limbic epilepsy has an advantage that this syndrome is strictly limited to a set of abnormalities that have a similar appearance among individuals (Bertram, 2008).

3.1. Why Do We Need Animal Models for Epilepsy?

Epileptic studies in humans are limited due to the heterogeneity of patient populations and the difficulty to study therapy related effects in a systematic way. Therefore, animal models are required to mimic epilepsy in humans. There are several available animal models of epilepsy including neurochemical agents, electrical stimulation, thermal or hypoxic insults, traumatic injuries, optogenetics, and rodent strains with idiopathic or audiogenic-induced seizures. Kandratavicius et al., have summarized important features, uses, and limitations of these animal models (see review (Kandratavicius et al., 2014)).

These animal models are useful to screen anti-epileptic drugs (AEDs). Most economic frequently used chemoconvulsants are useful, due to the fast examination of epileptogenic mechanisms and AED screening. However, there are certain drawbacks like, high mortality of subjects, high variability in the frequency and severity of spontaneous seizures. However, Electrical stimulation protocols are less harmful to the subjects and offer better control to induce seizures though electrophysiological procedures can be very costly and time-consuming. Nonchemical and nonelectrical insults might mimic clinical conditions of developmental epilepsies. These methods are inconsistent and an inconsistency in seizure susceptibility is depending on experimental procedures that hamper fair comparisons between rodents and humans.

The use of transgenic animals (seizure-prone mouse lines) have eliminated many of the experimentally induced seizures, but their genetic alterations are not fully identified and still required sensorial triggers to induce seizures in strains such as Wistar audiogenic rat (WARs) and genetic absence epilepsy in rats (GEAR) from Strasbourg. Such experimental models reflect the diversity of seizure types in humans. Even focal syndromes yield brain-wide alterations, thereby precipitating cognitive and psychiatric disturbances.

3.2. Epilepsy as Corrupt Neuronal Plasticity

Neuronal plasticity is a feature of nervous system, which is highly activity dependent and exquisite feature of nervous system to adapt to environmental changes. Synapse formation, elimination and remodeling to establish optimal connectivity, absolutely requires guidance by neural activity, which is highly synchronized between the neighboring neurons. When, this synchronized activity occurs in more than required level, generates epileptiform activity (Magnusson, 1993). Before going into details, it is important to understand two definitions, which helps to understand the role of neural plasticity in epileptogenesis. Epileptic mechanism responsible for seizure generation, displays excessive neuronal excitability leading disordered ictal activity (Avanzini et al., 2014). Epileptogenic mechanism responsible for epilepsy that enduring the tendency to generate epileptic seizures, dependent to hypothetical neurobiological processes leading to a permanent dysfunction of neuronal network (Avanzini et al., 2014). Clinical and experimental data showed that neuronal insult by trauma, infection initiate series of degenerative, regenerative and inflammatory changes, causing enduring epileptic aggregates. Latent period is associated with recurrent seizures and during this period biological change occurs, which may account for the progression of disease. These changes are not only structural (in hippocampus) but are also associated with alteration of neuronal excitability and synchronized activity of neurons, which comprises changes e.g., axonal sprouting, synaptic remodeling, gene and protein expression, gliosis, inflammation and angiogenesis leads epileptogenesis (Sutula, 2004). Some adaptation of neuronal plasticity may be contributing to development of epilepsy or its progression. As primary site of epileptogenesis sending the neuronal projections to other areas in brain, alterations in neuronal plasticity at primary site leads distant site epileptogenesis (Morrell and deToledo-Morrell, 1999). This could provide an explanation for the severity of epilepsy or propagation of epileptic activity to distant areas from primary site in some cases. Seizure experimental models provided evidences that seizure activity facilitate circuit specific neuroprotection (Kelly and McIntyre, 1994; Masco et al., 1999; Kondratyev et al., 2001; Penner et al., 2001) and nonspecific physiological responses to seizure cannot account for neuronal protection. Primarily limbic system after some injury with limited seizure activity (still adaptive and synchronizes neuronal firing) also can be neuroprotective. Moreover, in pathological conditions like epilepsy these rhythmic synchronized bursting is disturbed and repeatedly interrupting normal central nervous system (CNS) function. Avanzini et al., called this epileptogenic mechanism as a corruption of normal, adaptive neuroplasticity (Avanzini et al., 2014). It may happen due to lack of control on excitatory and inhibitory plasticity.

3.3. Imbalance in Hippocampal Excitatory and Inhibitory Plasticity in Epileptic Brain

Epilepsy stages are coinciding with the mechanism of long-term plasticity, learning and memory (Cain, 1989b; Meador, 2007), which is responsible to alter the factors that regulate the synaptic plasticity of excitatory and inhibitory circuits. There is a fine balance in excitatory glutamatergic and inhibitory GABAergic transmission, which defines the non-pathological function of brain. Alterations in synaptic plasticity causing the imbalance in excitatory/ inhibitory (E/I) components, make brain vulnerable to epilepsy. Kindling models of epilepsy showed that synaptic efficacy is changed according to the consistency of the electrical stimulus parameters and reception (Malenka and Bear, 2004). In hippocampal neurons, high frequency stimulation or concurrence of pre- and postsynaptic activity can increase the glutamatergic synaptic efficacy and can change GABAergic synaptic strength. Long term plasticity of excitatory synapses can be regulated by the strength of GABAergic inhibition. Moreover, previous research also showed that in CA1 pyramidal neurons, repetitive stimulation induce long term potentiation (LTP) at glutamatergic synapses can trigger long term depression (LTD) at GABAergic synapses (Chevaleyre and Castillo, 2003; Ahumada et al.,

2013). These studies suggested, abnormal alteration of long term plasticity, which require increased excitatory and decreased inhibitory transmission may contributes to the progression and consolidation of epilepsy (Cain, 1989a; Meador, 2007).

3.4. Differences in Glutamate Plasticity in the Epilepsy

The enhanced level of excitatory neurotransmitter "glutamate" identified as a key neuropathophysiological change in the epileptic brain. At the synapse the N-methyl-D-aspartate (NMDA) receptors required to maintain input and activity dependent temporal precision to promote protein synthesis, translocation of receptors and gene expression resulting long lasting synaptic plasticity via Ca^{2+} influx through voltage gated channels (Malenka, 2003). *In-vivo* (Goddard et al., 1969; Racine, 1972) and *in-vitro* (Stasheff et al., 1985; 1989) studies of chronic epilepsy models have shown that the spike bursts due to tetanus stimulation, whose duration is progressed with repetitive stimulation. Likewise, pharmacological models of epilepsy (pilocarpine and PTZ) displayed recurrent seizures in result of repetitive neuronal activation (Dhir, 2012; Mazzuferi et al., 2012; Lopes et al., 2013). In both cases, the hyperexcitability of neurons inhibited by NMDA receptor antagonists (i.e., APV and MK801). The overexpression of NMDA receptors (NMDARs) and α-amino-3-hydroxy-5-methyl-4-isoxazolepropionic acid receptors (AMPARs) have been documented in hippocampal formation in both, humans TLE patients and epilepsy animal models (Lopes et al., 2013; Naylor et al., 2013). Similarly, immunohistochemical studies have shown alteration in the glutamate receptor 1 (GluR1) and GluR2/3 or NR1 and NR2 subunits of AMPARs and NMDARs respectively, in the hippocampal subfields of TLE patients (Mathern et al., 1998). NR1was increased in CA3-CA1, while GluR2/3 were expressed in soma and dendrites in both proximal neurons and dentate granule cells (Lopes et al., 2013; Naylor et al., 2013). Therefore, glutamate receptors are main target since two decades to develop anti-epileptic drug (AEDs).

3.5. GABAergic Synaptic Plasticity Mechanism in Epilepsy

The GABAergic interneurons have a crucial role to synchronize local neuronal network and functional coupling in the brain (Mann and Paulsen, 2007). The activity dependent alterations in inhibitory neurotransmission are associated with alterations in GABAergic synapse efficacy, synaptic modifications, formation and eliminations of inhibitory contacts (Vithlani et al., 2011). Unlike, Glutamatergic synapses, depression of GABAergic synapses, could be proepileptogenic. At synapses $GABA_A$Rs constitutively undergo endocytosis, rapidly recycled and targeted for lysosomal degradation (Kittler et al., 2004). Studies in epileptic patients documented changes in the number and the expression of GABARs (Loup et al., 2009). Therefore, the association between the number of postsynaptic $GABA_A$Rs and the strength of the synapse is crucial for the cell stability (Nusser et al., 1997; 1998). However, $GABA_A$Rs number, recurrent membrane insertion of newly synthesized or recycled receptors is prerequisite to regulate neuronal plasticity (Loup et al., 2000; Naylor et al., 2005). In the status epilepticus (SE), the trafficking of $GABA_A$Rs is altered resulting prolonged seizures and responsible for the benzodiazepine pharmaco-resistance in the patients (Naylor et al., 2005; Goodkin et al., 2008). Eckel et al., have noticed a correlation between decrease $GABA_A$Rs, which alters intracellular calcium concentration and calcineurin activity, consequently contributing to pathophysiological signaling in SE (Eckel et al., 2014). Compensatory axonal sprouting is the main cause for the loss of GABAergic interneurons (reduction, restoration, and potentiation). Somatostatin (SOM) and cholecystokinin (CCK) are CB1 expressing GABAergic neurons in the hippocampus. They play a significant role in inhibitory synaptic transmission, synaptogenesis and form functional but aberrant circuitry in a mouse model of epilepsy (Peng et al., 2013; Liu et al., 2014). Recent rat model of TLE showed, a loss of CCK containing GABAergic terminals and synapses in the inner molecular layer of the dentate gyrus causes the reduction of CCK-containing GABAergic synaptic transmission to dentate gyrus cells (DGCs), tend to reduce seizure threshold (Sun et al., 2013).

3.6. Astrocyte and Neuroglia and Hippocampal Plasticity Change in Epilepsy

Astrocytes are playing several crucial roles in homeostasis of neuronal plasticity. In many cases, astrocytic fine protrusions are located adjacent to the excitatory synapses known as perisynaptic astrocytic processes (PAPs). Presence of PAPs near to synapses are subjective to physiological state, local neuronal activity and induction of synaptic plasticity. PAPs have leaf like ultrathin shape and have large surface to volume ratio, which may facilitate the local chemical compartmentalization (Rusakov et al., 2011). Astrocytes are making special functional integrated structure with synapses, called tripartite synapse. Tripartite synapse comprises presynaptic membrane, postsynaptic membrane and astrocyte/ glial protrusion, which produce chemical synaptic activity. PAPs are expressing numerous proteins related to efficient neuronal activity (Haydon, 2001; Perea et al., 2009). Additionally, PAPs also contain glutamine synthetase (Derouiche and Frotscher, 1991), metabotropic glutamate receptors (mGluR) (Tamaru et al., 2001; Lavialle et al., 2011; Arizono et al., 2012, Charles et al., 2003), aquaporins (Nagelhus et al., 2004; Thrane et al., 2010), Kir4.1 (Higashi et al., 2001; Nagelhus et al., 2004), glutamate transporters (Chaudhry et al., 1995; Haugeto et al., 1996; Lehre and Danbolt, 1998), actin-binding proteins (Lavialle et al., 2011; Molotkov et al., 2013) and cell adhesion molecules (Theodosis et al., 2002). Astrocytes are responsible for glutamate reuptake (Danbolt, 2001) and energy supply to neuronal synapses (Newman, 1993; Kimelberg, 2010; Suzuki et al., 2011). Beside that, astrocytes also release various signaling molecules (gliotransmitters) such as glutamate (Bezzi et al., 2004), D-serine (Panatier et al., 2006; Henneberger et al., 2010), adenosine triphosphate (ATP) (Cao et al., 2013), which could influence proximal synapses. Electron microscope data showed that astrocytes contain cellular organelle related to the Ca^{2+} signaling, near to excitatory synapses (Patrushev et al., 2013). Recent data also showed that astrocytic processes have Ca^{2+} store and sinks (Lovatt et al., 2007; Sahlender et al., 2014; Volterra et al., 2014). In hippocampus, there are two types of astrocytes, one expressing glutamate transporter and other have AMPARs (Matthias et al., 2003). Hippocampal astrocytes have high level of glutamate transporter GLT-1 (Chaudhry et al., 1995; Furuta et al., 1997; Tzingounis and Wadiche, 2007), which is dependent on neuronal activity (Yang et al., 2009; Benediktsson et al., 2011). In rat hippocampus approximately 62-90% neuronal spines are in contact with astrocytes and seem like that astrocytic processes approach preferentially perforated postsynaptic densities (Witcher et al., 2006) at postsynaptic site (Lehre and Rusakov, 2002). Whereas, PAPs are covering CA3 neuronal synapses, which is efficient to avoid spillover (Rollenhagen et al., 2007). Moreover, in CA1 region astrocytic processes barely cover axon-spine interphase and allow extrasynaptic escape of glutamate (Bernardinelli et al., 2014a). Studies showed that PAPs are motile and it is synapse specific. PAPs motility is dependent on neuronal activity, mGluR, Ca^{2+} and IP_3 (Perez-Alvarez et al., 2014; Bernardinelli et al., 2014b). Structural plasticity of PAPs may affect synaptic environment (Nishida and Okabe, 2007). Furthermore, neurons and astrocytes are forming perineuronal nets (PNNs) (Faissner et al., 2010) may play role in compartmentalization and prevent neurotransmitter spillover (Frischknecht and Seidenbecher, 2008; Frischknecht et al., 2009). Thus, astrocytes prevent the synaptic crosstalk by precluding the neurotransmitter spillover, which is brain region specific (OLIET et al., 2008; Lange et al., 2012; Bernardinelli et al., 2014a). Thus, astrocytes may able to change local environmental, limiting or allowing molecules, signal exchange at synapses and can modulate local excitatory circuits. Modified astrocytic coverage could affect synaptic efficacy by alteration in lactate transport (Suzuki et al., 2011).

3.7. Role of Calcium Channels in Synaptic Homeostasis and Epilepsy

There are various ion channels expressed on neuronal membrane and mutations in these channels play role in development or progression of neuropathophysiological conditions. Similarly, several reports showed, the contribution of different ion channels in epilepsy (see review for more information (Wolfart and Laker, 2015)). Although, in this chapter, we will focus only on Ca^{2+} channels and its contribution in homeostasis of synaptic plasticity and epilepsy. Research studies displayed that voltage gated calcium channels (Ca_V) primarily setting up or playing role to regulate neuronal activity by using pharmacological agents to block calcium channels or by using genetic models. These studies also shed light how indirectly or directly calcium channel may participate in epilepsy (Frank, 2014). However, it is still enigmatic, if calcium channel mutations could be the sole reason for the generation of epilepsy. Recent studies showed contribution of calcium channels in many processes, required for homeostatic synaptic plasticity (HSP). Whereas, HSP ensures that neuronal activity will be maintain within the physiological level, its alterations are dependent on presynaptic neurotransmitter release, neurotransmitter receptor expression, ion channel density and ion channel conductance (Pérez-Otaño and Ehlers, 2005; Davis, 2006; Marder and Goaillard, 2006; Turrigiano, 2008; Maffei and Fontanini, 2009; Pozo and Goda, 2010; Watt and Desai, 2010; Marder, 2012; Turrigiano, 2012). It is known that Ca_V playing crucial role in neurotransmission, gene transcription and intracellular signaling cascades (Catterall, 2000; Tedford and Zamponi, 2006; Catterall and Few, 2008; Dolphin, 2009; Currie, 2010; Cain and Snutch, 2011; Lipscombe et al., 2013). Ca_V channels contribute regulatory functions at many levels from regulation of channel subunit mRNA splicing, channel subunit trafficking (Lipscombe et al., 2013), targeting, cellular distribution of Ca_V channels (Herlitze et al., 2004) and G-protein mediated inhibition of Ca_V channel (Tedford and Zamponi, 2006; Currie, 2010; Zamponi and Currie, 2012). Calcium binding regulatory messengers such as calmodulin binds with Ca^{2+} to activate signaling cascade, which can inhibit or facilitate Ca_V channels (Halling et al., 2005; Dunlap, 2007; Liu et al., 2007; Catterall and Few, 2008; Minor and Findeisen, 2010; Christel and Lee, 2011). Ca_V channels mediate fast neurotransmission and Ca^{2+} influx by sensing membrane depolarization, which is sensed by synaptotagmin I (calcium sensing molecules) (Catterall et al., 2005; Striessnig, 2016) and triggers fast neurotransmitter release (Giraudo et al.,

2006; Schaub et al., 2006; Tang et al., 2006). Studies have suggested the role of Ca_V channels in HSP and its expression level is a major determinant of homeostasis of synaptic plasticity (Zhao et al., 2011). Presynaptic Ca_V channels are required for metaplasticity. Recent study showed that AMPA receptor blockade for long time induce increased vesical pool size and turnover rate due to lack of Ca^{2+} influx via Ca_V channels. It has been also seen that at postsynaptic membrane, GluR1 concentrated at largest synapses and overall synapse size is altered. This could be explained by metaplasticity (Zhao et al., 2011). Furthermore, other studies showed that blockade of AMPA receptors in absence of TTX inducing homeostatic responses by integrating of new GluA2-lacking AMPA receptors postsynaptically and a retrograde signaling process, increasing presynaptic release properties (Thiagarajan et al., 2005; Gong et al., 2007). Many evidences suggested, Ca_V channels are required for homeostatic regulation of synaptic plasticity (Magee, 1997; Hardingham et al., 1998; Weisskopf et al., 1999; Mermelstein et al., 2000; Brosenitsch and Katz, 2001; Dolmetsch et al., 2001; Pak and Sheng, 2003; Frank et al., 2009; Bliss, n.d.). Moreover, Ca_V channels are also contributing to stabilize excitatory-inhibitory connections by decreasing the abundance of inhibitory receptors at GABAergic synapses in chronic neuronal silencing (Kilman et al., 2002; Swanwick et al., 2005; Saliba et al., 2007; 2009). Optogenetics results illustrated that in response to the excitatory stimulation, both NMDR and AMPAR mediated responses are downregulated at postsynaptic site. This form of homeostatic depression is caused by L-type Ca_V channel (Goold and Nicoll, 2010). Mutational studies provided evidences that mutations in Ca_V channels (Ca_V 2.1, Ca_V 3.1, Ca_V 3.2) contribute to absence seizure, idiopathic generalized epilepsy (IGE) susceptibility and spontaneous recurrent seizures (Letts et al., 1998; Barclay et al., 2001; Letts et al., 2003; Becker et al., 2008). Furthermore, Ca_V3 channels are also expressed at midline thalamic nuclei (Talley et al., 1999), a part of epilepsy network (Bertram et al., 2007). The reuniens nucleus (RE) sends excitatory projections to CA1 dendritic region (Wouterlood et al., 1990) and entorhinal cortex (Vertes et al., 2006). Bertram et al., has illustrated that midline thalamus is an important part of TLE circuit (Bertram et al., 2001;

Bertram, 2008). Graef et al., has used a computational model for burst discharges of thalamic neurons to understand alterations in epileptogenesis by decease or increase the tendency of burst firing in seizure free condition or during spontaneous recurrent seizures (SRS), respectively (Graef et al., 2009). Thus, Ca_V3 channel current play crucial role to neuronal stability (Llinas, 1988; Errington et al., 2010).

4. Drug Targets and Current AEDs

All classical used antiepileptic drugs exert their effects first, by reducing glutamatergic transmission and neuronal excitability like levetiracetam, oxcarbazepine and lamotrigine (Landmark, 2006; Stief et al., 2007) or secondly, by increasing GABAergic inhibition drugs like vigabatrin, tiagabine and valproate. Commonly used drug Levetiracetam mainly binds selectively to proteins of neurotransmitter release machinery and reducing glutamatergic transmission (Sugaya et al., 2010). Several other new AEDs are accessible to treat epilepsy, most of them block voltage-gated sodium or calcium channels, enhance GABA function by activation or by positive allosteric modulation of $GABA_A$ receptors, inhibit GABA aminotransferase, or inhibit GABA reuptake from the synaptic cleft (Klitgaard et al., 2013; Serrano and Kanner, 2015). However, drug-resistance is not uncommon for the treatment of distinct type of epilepsies. Therefore, specific mode of action for the treatment of distinct epilepsies is required to develop new treatment options (Steinhoff, 2015). Current AEDs advancement occur in the direction of inhibiting excitatory neurotransmission to prevent epileptogenesis and seizure spread (Rogawski, 2011). Several research groups found that drugs acting through noncompetitive (negative allosteric) mechanism to inhibit AMPA receptors with less side effects may solve the purpose like; GYKI 53655 (GYKI) (Donevan et al., 1994; Balannik et al., 2005), CP 465022 (CP) (Menniti et al., 2000; Lazzaro et al., 2002; Balannik et al., 2005) and the pyridone perampane (PMP) (Bialer et al., 2010; Hibi et al., 2012; Chen et al., 2014). However, only PMP is considered medically innocuous and effective AED with low dose and with less adverse side effects (Patsalos, 2014; Steinhoff et al., 2014; Steinhoff, 2015). Though, at higher doses, patients do experience side effects, including somnolence, dizziness, fatigue, irritability, nausea, headache, and falls, as well as depression and aggression (Coyle et al., 2014; Rugg-Gunn, 2014; Steinhoff et al., 2014), indicating the requirement of safer and more efficacious drugs. Recently, a novel AMPA receptor noncompetitive inhibitory binding site has been identified, preventing conformation change that leads to the ion channel opening (Yelshanskaya et al., 2016). Accordingly, the noncompetitive inhibitor binding pocket represents a new promising target to design drug with enhanced safety measurements and efficacy.

Apart from $GABA_A$ receptors, the $GABA_B$ receptors persistent activation in epileptic mice can reduce inhibitory output from hippocampal cholecystokinin basket cell interneurons, which lead to disinhibition in hippocampal networks, enhance gamma activity and promote pathological hyperexcitability. The current data in this direction suggested the role of $GABA_B$ to control epileptic activity and may provide therapeutic target for the treatment of seizures (Dugladze et al., 2013).

CONCLUSION

It seems that homeostasis regulation is playing critical role to maintain sustainable physiological neuronal activity, which is examined by various monitors. These monitors are located either in soma or near to the synaptic densities where they keep tight control to provide relevant signal to presynaptic or postsynaptic neurons to maintain neuronal activity. Homeostasis regulation is dependent on wide range of factors from ionotropic channels to astrocyte released trophic factors. It is hypothesized that epilepsy is initiated by defects in neuronal homeostatic signaling, which could define as instable neuronal function. Stability in neuronal homeostasis is contributed by excitatory and inhibitory balance between excitatory and inhibitory synapses that is established during the development and preserved throughout the life.

REFERENCES

Abraham WC, Bear MF (1996) Metaplasticity: the plasticity of synaptic plasticity. *Trends Neurosci* 19:126.

Ahumada J, de Sevilla DF, Couve A, Buño W, Fuenzalida M (2013) Long-term depression of inhibitory synaptic transmission induced by spike-timing dependent plasticity requires coactivation of endocannabinoid and muscarinic receptors. *Hippocampus* 23:1439–1452.

Alger BE (2005) Endocannabinoids and their implications for epilepsy. *Epilepsy Curr* 4:169–173.

Arizono M, Bannai H, Nakamura K, Niwa F, Enomoto M, Matsu-Ura T, Miyamoto A, Sherwood MW, Nakamura T, Mikoshiba K (2012) Receptor-selective diffusion barrier enhances sensitivity of astrocytic processes to metabotropic glutamate receptor stimulation. *Sci Signal* 5:ra27.

Avanzini G, Forcelli PA, Gale K (2014) Are there really "epileptogenic" mechanisms or only corruptions of "normal" plasticity? *Adv Exp Med Biol* 813:95–107.

Balannik V, Menniti FS, Paternain AV, Lerma J, Stern-Bach Y (2005) Molecular mechanism of AMPA receptor noncompetitive antagonism. *Neuron* 48:279–288.

Barclay J, Balaguero N, Mione M, Ackerman SL, Letts VA, Brodbeck J, Canti C, Meir A, Page KM, Kusumi K, Perez-Reyes E, Lander ES, Frankel WN, Gardiner RM, Dolphin AC, Rees M (2001) Ducky mouse phenotype of epilepsy and ataxia is associated with mutations in the Cacna2d2 gene and decreased calcium channel current in cerebellar Purkinje cells. *J Neurosci* 21:6095–6104.

Bateup HS, Johnson CA, Denefrio CL, Saulnier JL, Kornacker K, Sabatini BL (2013) Excitatory/inhibitory synaptic imbalance leads to hippocampal hyperexcitability in mouse models of tuberous sclerosis. *Neuron* 78:510–522.

Becker AJ, Pitsch J, Sochivko D, Opitz T, Staniek M, Chen C-C, Campbell KP, Schoch S, Yaari Y, Beck H (2008) Transcriptional upregulation of

Cav3.2 mediates epileptogenesis in the pilocarpine model of epilepsy. *J Neurosci* 28:13341–13353.

Benediktsson AM, Marrs GS, Tu JC, Worley PF, Rothstein JD, Bergles DE, Dailey ME (2011) Neuronal activity regulates glutamate transporter dynamics in developing astrocytes. *Glia* 60:175–188.

Bernardinelli Y, Muller D, Nikonenko I (2014a) Astrocyte-synapse structural plasticity. *Neural Plast* 2014:232105.

Bernardinelli Y, Randall J, Janett E, Nikonenko I, König S, Jones EV, Flores CE, Murai KK, Bochet CG, Holtmaat A, Muller D (2014b) Activity-dependent structural plasticity of perisynaptic astrocytic domains promotes excitatory synapse stability. *Curr Biol* 24:1679–1688.

Bertram EH (2008) Temporal lobe epilepsy: where do the seizures really begin? *Epilepsy Behav* 14 Suppl 1:32–37.

Bertram EH, Mangan PS, Zhang D, Scott CA, Williamson JM (2001) The Midline Thalamus: Alterations and a Potential Role in Limbic Epilepsy. *Epilepsia* 42:967.

Bertram EH, Zhang D, Williamson JM (2007) Multiple roles of midline dorsal thalamic nuclei in induction and spread of limbic seizures. *Epilepsia* 49:256–268.

Bezzi P, Gundersen V, Galbete JL, Seifert G, Steinhäuser C, Pilati E, Volterra A (2004) Astrocytes contain a vesicular compartment that is competent for regulated exocytosis of glutamate. *Nat Neurosci* 7:613–620.

Bi GQ, Poo MM (1998) Synaptic modifications in cultured hippocampal neurons: dependence on spike timing, synaptic strength, and postsynaptic cell type. *J Neurosci* 18:10464–10472.

Bialer M, Johannessen SI, Levy RH, Perucca E, Tomson T, White HS (2010) Progress report on new antiepileptic drugs: a summary of the Tenth Eilat Conference (EILAT X). *Epilepsy Res* 92:89–124.

Bito H, Deisseroth K, Tsien RW (1996) CREB Phosphorylation and Dephosphorylation: A Ca2+- and Stimulus Duration–Dependent Switch for Hippocampal Gene Expression. *Cell* 87:1203.

Blair RE, Deshpande LS, Sombati S, Falenski KW, Martin BR, DeLorenzo RJ (2006) Activation of the cannabinoid type-1 receptor mediates the anticonvulsant properties of cannabinoids in the hippocampal neuronal culture models of acquired epilepsy and status epilepticus. *J Pharmacol Exp Ther* 317:1072–1078.

Bliss T (n.d.) Faculty of 1000 evaluation for Depolarization-induced long-term depression at hippocampal mossy fiber-CA3 pyramidal neuron synapses. *Cell.*

Bliss TV, Collingridge GL (1993) A synaptic model of memory: long-term potentiation in the hippocampus. *Nature* 361:31–39.

Boison D (2011) Adenosine dysfunction in epilepsy. *Glia* 60:1234–1243.

Boison D (2013) Adenosine kinase: exploitation for therapeutic gain. *Pharmacol Rev* 65:906–943.

Bojnik E, Turunç E, Armağan G, Kanıt L, Benyhe S, Yalçın A, Borsodi A (2011) Changes in the cannabinoid (CB1) receptor expression level and G-protein activation in kainic acid induced seizures. *Epilepsy Res* 99:64–68.

Braun AP, Schulman H (1995) The multifunctional calcium/calmodulin-dependent protein kinase: from form to function. *Annu Rev Physiol* 57:417–445.

Bredt DS (2000) Cell biology. Reeling CASK into the nucleus. *Nature* 404:241–242.

Broadie K (n.d.) Faculty of 1000 evaluation for Neuroligin expressed in nonneuronal cells triggers presynaptic development in contacting axons. *Cell.*

Brocke L, Srinivasan M, Schulman H (1995) Developmental and regional expression of multifunctional Ca2+/calmodulin-dependent protein kinase isoforms in rat brain. *J Neurosci* 15:6797–6808.

Brosenitsch TA, Katz DM (2001) Physiological patterns of electrical stimulation can induce neuronal gene expression by activating N-type calcium channels. *J Neurosci* 21:2571–2579.

Butz S, Okamoto M, Südhof TC (1998) A Tripartite Protein Complex with the Potential to Couple Synaptic Vesicle Exocytosis to Cell Adhesion in Brain. *Cell* 94:773.

Cain DP (1989a) Long-term potentiation and kindling: how similar are the mechanisms? *Trends in neurosciences* 12:6–10.

Cain SM, Snutch TP (2011) Voltage-gated calcium channels and disease. *Biofactors* 37:197–205.

Cao X, Li L-P, Wang Q, Wu Q, Hu H-H, Zhang M, Fang Y-Y, Zhang J, Li S-J, Xiong W-C, Yan H-C, Gao Y-B, Liu J-H, Li X-W, Sun L-R, Zeng Y-N, Zhu X-H, Gao T-M (2013) Astrocyte-derived ATP modulates depressive-like behaviors. *Nat Med* 19:773–777.

Carroll RC, Beattie EC, Xia H, Lüscher C, Altschuler Y, Nicoll RA, Malenka RC, Zastrow von M (1999) Dynamin-dependent endocytosis of ionotropic glutamate receptors. *Proc Natl Acad Sci USA* 96:14112.

Catterall WA (2000) *Structure and regulation of voltage-gated Ca2+ channels.*

Catterall WA, Few AP (2008) Calcium channel regulation and presynaptic plasticity. *Neuron* 59:882–901.

Catterall WA, Perez-Reyes E, Snutch TP, Striessnig J (2005) International Union of Pharmacology. XLVIII. Nomenclature and structure-function relationships of voltage-gated calcium channels. *Pharmacol Rev* 57:411–425.

Charles KJ, Calver AR, Jourdain S, Pangalos MN (2003) Distribution of a GABAB-like receptor protein in the rat central nervous system. *Brain Res* 989:135–146.

Chaudhry FA, Lehre KP, van Lookeren Campagne M, Ottersen OP, Danbolt NC, Storm-Mathisen J (1995) Glutamate transporters in glial plasma membranes: Highly differentiated localizations revealed by quantitative ultrastructural immunocytochemistry. *Neuron* 15:711.

Chemin J, Monteil A, Perez-Reyes E, Nargeot J, Lory P (2001) Direct inhibition of T-type calcium channels by the endogenous cannabinoid anandamide. *EMBO J* 20:7033–7040.

Chen C-Y, Matt L, Hell JW, Rogawski MA (2014) Perampanel inhibition of AMPA receptor currents in cultured hippocampal neurons. *PLoS ONE* 9:e108021.

Chen K, Neu A, Howard AL, Földy C, Echegoyen J, Hilgenberg L, Smith M, Mackie K, Soltesz I (2007) Prevention of plasticity of endocannabinoid signaling inhibits persistent limbic hyperexcitability caused by developmental seizures. *J Neurosci* 27:46–58.

Chen K, Ratzliff A, Hilgenberg L, Gulyás A, Freund TF, Smith M, Dinh TP, Piomelli D, Mackie K, Soltesz I (2003) Long-Term Plasticity of Endocannabinoid Signaling Induced by Developmental Febrile Seizures. *Neuron* 39:599.

Chevaleyre V, Castillo PE (2003) Heterosynaptic LTD of hippocampal GABAergic synapses: a novel role of endocannabinoids in regulating excitability. *Cell* 38:461–472.

Christel C, Lee A (2011) Ca2+-dependent modulation of voltage-gated Ca2+ channels. *Biochim Biophys Acta* 1820:1243–1252.

Coyle H, Clough P, Cooper P, Mohanraj R (2014) Clinical experience with perampanel: focus on psychiatric adverse effects. *Epilepsy Behav* 41:193–196.

Crabtree GR (1999) Generic Signals and Specific Outcomes. *Cell* 96:611.

Currie KPM (2010) G protein modulation of CaV2 voltage-gated calcium channels. *Channels (Austin)* 4:497–509.

Danbolt NC (2001) Glutamate uptake. *Prog Neurobiol* 65:1.

Davis GW (2006) Homeostatic control of neural activity: from phenomenology to molecular design. *Annu Rev Neurosci* 29:307–323.

Davis GW (2013) Homeostatic signaling and the stabilization of neural function. *Neuron* 80:718–728.

Davis GW, Bezprozvanny I (2001) Maintaining the stability of neural function: a homeostatic hypothesis. *Annu Rev Physiol* 63:847–869.

Davis GW, DiAntonio A, Petersen SA, Goodman CS (1998) Postsynaptic PKA Controls Quantal Size and Reveals a Retrograde Signal that Regulates Presynaptic Transmitter Release in Drosophila. *Neuron* 20:305.

Davis GW, Goodman CS (1998a) Synapse-specific control of synaptic efficacy at the terminals of a single neuron. *Nature* 392:82–86.

Davis GW, Goodman CS (1998b) Genetic analysis of synaptic development and plasticity: homeostatic regulation of synaptic efficacy. *Current opinion in neurobiology* 8:149–156.

Davis GW, Schuster CM, Goodman CS (1997) Genetic Analysis of the Mechanisms Controlling Target Selection: Target-Derived Fasciclin II Regulates the Pattern of Synapse Formation. *Neuron* 19:561.

De Koninck P (1998) Sensitivity of CaM Kinase II to the Frequency of Ca2+ Oscillations. *Science* 279:227.

Deisseroth K, Bito H, Tsien RW (1996) Signaling from Synapse to Nucleus: Postsynaptic CREB Phosphorylation during Multiple Forms of Hippocampal Synaptic Plasticity. *Neuron* 16:89.

Deisseroth K, Heist EK, Tsien RW (1998) Translocation of calmodulin to the nucleus supports CREB phosphorylation in hippocampal neurons. *Nature* 392:198–202.

Derouiche A, Frotscher M (1991) Astroglial processes around identified glutamatergic synapses contain glutamine synthetase: evidence for transmitter degradation. *Brain Res* 552:346–350.

Desai NS, Rutherford LC, Turrigiano GG (1999) Plasticity in the intrinsic excitability of cortical pyramidal neurons. *Nat Neurosci* 2:515–520.

Dhir A (2012) Pentylenetetrazol (PTZ) kindling model of epilepsy. *Curr Protoc Neurosci Chapter* 9:Unit9.37.

Dias RB, Rombo DM, Ribeiro JA, Henley JM, Sebastião AM (2013) Adenosine: setting the stage for plasticity. *Trends in neurosciences* 36:248–257.

Dolmetsch RE, Pajvani U, Fife K, Spotts JM, Greenberg ME (2001) Signaling to the nucleus by an L-type calcium channel-calmodulin complex through the MAP kinase pathway. *Science* 294:333–339.

Dolphin AC (2009) Calcium channel diversity: multiple roles of calcium channel subunits. *Current opinion in neurobiology* 19:237–244.

Domenici MR, Azad SC, Marsicano G, Schierloh A, Wotjak CT, Dodt H-U, Zieglgänsberger W, Lutz B, Rammes G (2006) Cannabinoid receptor type 1 located on presynaptic terminals of principal neurons in the forebrain controls glutamatergic synaptic transmission. *J Neurosci* 26:5794–5799.

Donevan SD, Yamaguchi S, Rogawski MA (1994) Non-N-methyl-D-aspartate receptor antagonism by 3-N-substituted 2,3-benzodiazepines:

relationship to anticonvulsant activity. *J Pharmacol Exp Ther* 271:25–29.

Dugladze T, Maziashvili N, Börgers C, Gurgenidze S, Häussler U, Winkelmann A, Haas CA, Meier JC, Vida I, Kopell NJ, Gloveli T (2013) GABA(B) autoreceptor-mediated cell type-specific reduction of inhibition in epileptic mice. *Proc Natl Acad Sci USA* 110:15073–15078.

Dunlap K (2007) Calcium channels are models of self-control. *J Gen Physiol* 129:379–383.

Eckel R, Szulc B, Walker MC, Kittler JT (2014) Activation of calcineurin underlies altered trafficking of α2 subunit containing GABAA receptors during prolonged epileptiform activity. *Neuropharmacology* 88:82–90.

Eichler SA, Kirischuk S, Jüttner R, Schaefermeier PK, Schafermeier PK, Legendre P, Lehmann T-N, Gloveli T, Grantyn R, Meier JC (2009) Glycinergic tonic inhibition of hippocampal neurons with depolarizing GABAergic transmission elicits histopathological signs of temporal lobe epilepsy. *J Cell Mol Med* 12:2848–2866.

Eichler SA, Meier JC (2008) E-I balance and human diseases - from molecules to networking. *Front Mol Neurosci* 1:2.

Errington AC, Renger JJ, Uebele VN, Crunelli V (2010) State-dependent firing determines intrinsic dendritic Ca2+ signaling in thalamocortical neurons. *J Neurosci* 30:14843–14853.

Faissner A, Pyka M, Geissler M, Sobik T, Frischknecht R, Gundelfinger ED, Seidenbecher C (2010) Contributions of astrocytes to synapse formation and maturation - Potential functions of the perisynaptic extracellular matrix. *Brain Res Rev* 63:26–38.

Frank CA (2013) Homeostatic plasticity at the Drosophila neuromuscular junction. *Neuropharmacology* 78:63–74.

Frank CA (2014) How voltage-gated calcium channels gate forms of homeostatic synaptic plasticity. *Front Cell Neurosci* 8:40.

Frank CA, Kennedy MJ, Goold CP, Marek KW, Davis GW (2006) Mechanisms underlying the rapid induction and sustained expression of synaptic homeostasis. *Neuron* 52:663–677.

Frank CA, Pielage J, Davis GW (2009) A presynaptic homeostatic signaling system composed of the Eph receptor, ephexin, Cdc42, and CaV2.1 calcium channels. *Neuron* 61:556–569.

Frischknecht R, Heine M, Perrais D, Seidenbecher CI, Choquet D, Gundelfinger ED (2009) Brain extracellular matrix affects AMPA receptor lateral mobility and short-term synaptic plasticity. *Nat Neurosci* 12:897–904.

Frischknecht R, Seidenbecher CI (2008) The crosstalk of hyaluronan-based extracellular matrix and synapses. *Neuron Glia Biol* 4:249–257.

Furuta A, Rothstein JD, Martin LJ (1997) Glutamate transporter protein subtypes are expressed differentially during rat CNS development. *J Neurosci* 17:8363–8375.

Giagtzoglou N, Lin YQ, Haueter C, Bellen HJ (2009) Importin 13 regulates neurotransmitter release at the Drosophila neuromuscular junction. *J Neurosci* 29:5628–5639.

Giraudo CG, Eng WS, Melia TJ, Rothman JE (2006) A clamping mechanism involved in SNARE-dependent exocytosis. *Science* 313:676–680.

Goddard GV, McIntyre DC, Leech CK (1969) A permanent change in brain function resulting from daily electrical stimulation. *Exp Neurol* 25:295–330.

Goffin K, Van Paesschen W, Van Laere K (2011) In vivo activation of endocannabinoid system in temporal lobe epilepsy with hippocampal sclerosis. *Brain* 134:1033–1040.

Gong B, Wang H, Gu S, Heximer SP, Zhuo M (2007) Genetic evidence for the requirement of adenylyl cyclase 1 in synaptic scaling of forebrain cortical neurons. *Eur J Neurosci* 26:275–288.

Goodkin HP, Joshi S, Mtchedlishvili Z, Brar J, Kapur J (2008) Subunit-specific trafficking of GABA(A) receptors during status epilepticus. *J Neurosci* 28:2527–2538.

Goold CP, Nicoll RA (2010) Single-cell optogenetic excitation drives homeostatic synaptic depression. *Neuron* 68:512–528.

Graef IA, Mermelstein PG, Stankunas K, Neilson JR, Deisseroth K, Tsien RW, Crabtree GR (1999) L-type calcium channels and GSK-3 regulate the activity of NF-ATc4 in hippocampal neurons. *Nature* 401:703–708.

Graef JD, Nordskog BK, Wiggins WF, Godwin DW (2009) An acquired channelopathy involving thalamic T-type Ca2+ channels after status epilepticus. *J Neurosci* 29:4430–4441.

Grubb MS, Burrone J (2010) Activity-dependent relocation of the axon initial segment fine-tunes neuronal excitability. *Nature* 465:1070–1074.

Gu Y, Barry J, McDougel R, Terman D, Gu C (2011) Alternative splicing regulates kv3.1 polarized targeting to adjust maximal spiking frequency. *J Biol Chem* 287:1755–1769.

Halling DB, Aracena-Parks P, Hamilton SL (2005) Regulation of Voltage-Gated Ca2+ Channels by Calmodulin. *Sci Signal* 2005:re15.

Hardingham GE, Cruzalegui FH, Chawla S, Bading H (1998) Mechanisms controlling gene expression by nuclear calcium signals. *Cell Calcium* 23:131.

Hata Y, Butz S, Südhof TC (1996) CASK: a novel dlg/PSD95 homolog with an N-terminal calmodulin-dependent protein kinase domain identified by interaction with neurexins. *J Neurosci* 16:2488–2494.

Haugeto O, Ullensvang K, Levy LM, Chaudhry FA, Honoré T, Nielsen M, Lehre KP, Danbolt NC (1996) Brain Glutamate Transporter Proteins Form Homomultimers. *J Biol Chem* 271:27715.

Hayashi Y (2000) Driving AMPA Receptors into Synapses by LTP and CaMKII: Requirement for GluR1 and PDZ Domain Interaction. *Science* 287:2262.

Haydon PG (2001) GLIA: listening and talking to the synapse. *Nat Rev Neurosci 2*:185–193.

Henneberger C, Papouin T, Oliet SHR, Rusakov DA (2010) Long-term potentiation depends on release of D-serine from astrocytes. *Nature* 463:232–236.

Henry FE, McCartney AJ, Neely R, Perez AS, Carruthers CJL, Stuenkel EL, Inoki K, Sutton MA (2012) Retrograde changes in presynaptic function driven by dendritic mTORC1. *J Neurosci* 32:17128–17142.

Herlitze S, Xie M, Han J, Hümmer A, Melnik-Martinez KV, Moreno RL, Mark MD (2004) Targeting mechanisms of high voltage-activated Ca2+ channels. *J Bioenerg Biomembr* 35:621–637.

Hibi S, Ueno K, Nagato S, Kawano K, Ito K, Norimine Y, Takenaka O, Hanada T, Yonaga M (2012) Discovery of 2-(2-oxo-1-phenyl-5-pyridin-2-yl-1,2-dihydropyridin-3-yl)benzonitrile (perampanel): a novel, noncompetitive α-amino-3-hydroxy-5-methyl-4-isoxazolepropanoic acid (AMPA) receptor antagonist. *J Med Chem* 55:10584–10600.

Higashi K, Fujita A, Inanobe A, Tanemoto M, Doi K, Kubo T, Kurachi Y (2001) An inwardly rectifying K(+) channel, Kir4.1, expressed in astrocytes surrounds synapses and blood vessels in brain. *Am J Physiol, Cell Physiol* 281:C922–931.

Hill AJ, Williams CM, Whalley BJ, Stephens GJ (2011) Phytocannabinoids as novel therapeutic agents in CNS disorders. *Pharmacol Ther* 133:79–97.

Irie M, Hata Y, Takeuchi M, Ichtchenko K, Toyoda A, Hirao K, Takai Y, Rosahl TW, Südhof TC (1997) Binding of neuroligins to PSD-95. *Science* 277:1511–1515.

Jonas P (2000) Dynamic Control of Presynaptic Ca2+ Inflow by Fast-Inactivating K+ Channels in Hippocampal Mossy Fiber Boutons. *Neuron* 28:927.

Kandratavicius L, Balista PA, Lopes-Aguiar C, Ruggiero RN, Umeoka EH, Garcia-Cairasco N, Bueno-Junior LS, Leite JP (2014) Animal models of epilepsy: use and limitations. *Neuropsychiatr Dis* Treat 10:1693–1705.

Karlócai MR, Tóth K, Watanabe M, Ledent C, Juhász G, Freund TF, Maglóczky Z (2011) Redistribution of CB1 cannabinoid receptors in the acute and chronic phases of pilocarpine-induced epilepsy. *PLoS ONE* 6:e27196.

Karmarkar UR, Buonomano DV (2006) Different forms of homeostatic plasticity are engaged with distinct temporal profiles. *Eur J Neurosci* 23:1575–1584.

Katona I, Freund TF (2008) Endocannabinoid signaling as a synaptic circuit breaker in neurological disease. *Nat Med* 14:923–930.

Katz LC, Shatz CJ (1996) Synaptic Activity and the Construction of Cortical Circuits. *Science* 274:1133.

Kawamura Y, Fukaya M, Maejima T, Yoshida T, Miura E, Watanabe M, Ohno-Shosaku T, Kano M (2006) The CB1 cannabinoid receptor is the major cannabinoid receptor at excitatory presynaptic sites in the hippocampus and cerebellum. *J Neurosci* 26:2991–3001.

Kelly ME, McIntyre DC (1994) Hippocampal kindling protects several structures from the neuronal damage resulting from kainic acid-induced status epilepticus. *Brain Res* 634:245–256.

Kilman V, van Rossum MCW, Turrigiano GG (2002) Activity deprivation reduces miniature IPSC amplitude by decreasing the number of postsynaptic GABA(A) receptors clustered at neocortical synapses. *J Neurosci* 22:1328–1337.

Kim SH, Ryan TA (2010) CDK5 serves as a major control point in neurotransmitter release. *Neuron* 67:797–809.

Kimelberg HK (2010) Functions of mature mammalian astrocytes: a current view. *Neuroscientist* 16:79–106.

Kirchheim F, Tinnes S, Haas CA, Stegen M, Wolfart J (2013) Regulation of action potential delays via voltage-gated potassium Kv1.1 channels in dentate granule cells during hippocampal epilepsy. *Front Cell Neurosci* 7:248.

Kittler JT, Arancibia-Carcamo IL, Moss SJ (2004) Association of GRIP1 with a GABA(A) receptor associated protein suggests a role for GRIP1 at inhibitory synapses. *Biochem Pharmacol* 68:1649–1654.

Kletke O, Sergeeva OA, Lorenz P, Oberland S, Meier JC, Hatt H, Gisselmann G (2013) New insights in endogenous modulation of ligand-gated ion channels: histamine is an inverse agonist at strychnine sensitive glycine receptors. *Eur J Pharmacol* 710:59–66.

Klitgaard H, Löscher W, Klitgaard H, Twyman RE, Schmidt D, Twyman RE, Schmidt D (2013) New avenues for anti-epileptic drug discovery and development. *Nat Rev Drug Discov* 12:757–776.

Kondratyev A, Sahibzada N, Gale K (2001) Electroconvulsive shock exposure prevents neuronal apoptosis after kainic acid-evoked status epilepticus. *Brain Res Mol Brain Res* 91:1–13.

Lai MM, Hong JJ, Ruggiero AM, Burnett PE, Slepnev VI, De Camilli P, Snyder SH (1999) The Calcineurin-Dynamin 1 Complex as a Calcium Sensor for Synaptic Vesicle Endocytosis. *J Biol Chem* 274:25963.

Landmark CJ (2006) Targets for antiepileptic drugs in the synapse. *Med Sci Monit* 13:RA1–7.

Lange SC, Bak LK, Waagepetersen HS, Schousboe A, Norenberg MD (2012) Primary cultures of astrocytes: their value in understanding astrocytes in health and disease. *Neurochem Res* 37:2569–2588.

Laplante M, Sabatini DM (2012) mTOR signaling in growth control and disease. *Cell* 149:274–293.

Lavialle M, Aumann G, Anlauf E, Pröls F, Arpin M, Derouiche A (2011) Structural plasticity of perisynaptic astrocyte processes involves ezrin and metabotropic glutamate receptors. *Proc Natl Acad Sci USA* 108:12915–12919.

Lawrence JJ, McBain CJ (2003) Interneuron diversity series: containing the detonation--feedforward inhibition in the CA3 hippocampus. *Trends in neurosciences* 26:631–640.

Lazzaro JT, Paternain AV, Lerma J, Chenard BL, Ewing FE, Huang J, Welch WM, Ganong AH, Menniti FS (2002) Functional characterization of CP-465,022, a selective, noncompetitive AMPA receptor antagonist. *Neuropharmacology* 42:143–153.

Legendre P, Förstera B, Jüttner R, Meier JC (2009) Glycine Receptors Caught between Genome and Proteome - Functional Implications of RNA Editing and Splicing. *Front Mol Neurosci* 2:23.

Lehre KP, Danbolt NC (1998) The number of glutamate transporter subtype molecules at glutamatergic synapses: chemical and stereological quantification in young adult rat brain. *J Neurosci* 18:8751–8757.

Lehre KP, Rusakov DA (2002) Asymmetry of glia near central synapses favors presynaptically directed glutamate escape. *Biophys J* 83:125–134.

Letts VA, Felix R, Biddlecome GH, Arikkath J, Mahaffey CL, Valenzuela A, Bartlett FS, Mori Y, Campbell KP, Frankel WN (1998) The mouse stargazer gene encodes a neuronal Ca2+-channel gamma subunit. *Nat Genet* 19:340–347.

Letts VA, Kang M-G, Mahaffey CL, Beyer B, Tenbrink H, Campbell KP, Frankel WN (2003) Phenotypic heterogeneity in the stargazin allelic series. *Mamm Genome* 14:506–513.

Lipscombe D, Allen SE, Toro CP (2013) Control of neuronal voltage-gated calcium ion channels from RNA to protein. *Trends in neurosciences* 36:598–609.

Liu G, Tsien RW (1995) Properties of synaptic transmission at single hippocampal synaptic boutons. *Nature* 375:404–408.

Liu Q, Chen B, Ge Q, Wang Z-W (2007) Presynaptic Ca2+/calmodulin-dependent protein kinase II modulates neurotransmitter release by activating BK channels at Caenorhabditis elegans neuromuscular junction. *J Neurosci* 27:10404–10413.

Liu Y-Q, Yu F, Liu W-H, He X-H, Peng B-W (2014) Dysfunction of hippocampal interneurons in epilepsy. *Neurosci Bull* 30:985–998.

Llinas R (1988) The intrinsic electrophysiological properties of mammalian neurons: insights into central nervous system function. *Science* 242:1654.

Lopes MW, Soares FMS, de Mello N, Nunes JC, Cajado AG, de Brito D, de Cordova FM, da Cunha RMS, Walz R, Leal RB (2013) Time-dependent modulation of AMPA receptor phosphorylation and mRNA expression of NMDA receptors and glial glutamate transporters in the rat hippocampus and cerebral cortex in a pilocarpine model of epilepsy. *Exp Brain Res* 226:153–163.

Loup F, Picard F, Yonekawa Y, Wieser H-G, Fritschy J-M (2009) Selective changes in GABAA receptor subtypes in white matter neurons of patients with focal epilepsy. *Brain* 132:2449–2463.

Loup F, Wieser HG, Yonekawa Y, Aguzzi A, Fritschy JM (2000) Selective alterations in GABAA receptor subtypes in human temporal lobe epilepsy. *J Neurosci* 20:5401–5419.

Lovatt D, Sonnewald U, Waagepetersen HS, Schousboe A, He W, Lin JH-C, Han X, Takano T, Wang S, Sim FJ, Goldman SA, Nedergaard M (2007) The transcriptome and metabolic gene signature of protoplasmic astrocytes in the adult murine cortex. *J Neurosci* 27:12255–12266.

Lu YM, Mansuy IM, Kandel ER, Roder J (2000) Calcineurin-Mediated LTD of GABAergic Inhibition Underlies the Increased Excitability of CA1 Neurons Associated with LTP. *Neuron* 26:197.

Lüscher C, Xia H, Beattie EC, Carroll RC, Zastrow von M, Malenka RC, Nicoll RA (1999) Role of AMPA Receptor Cycling in Synaptic Transmission and Plasticity. *Neuron* 24:649.

Mackie K, Lai Y, Westenbroek R, Mitchell R (1995) Cannabinoids activate an inwardly rectifying potassium conductance and inhibit Q-type calcium currents in AtT20 cells transfected with rat brain cannabinoid receptor. *J Neurosci* 15:6552–6561.

Maffei A, Fontanini A (2009) Network homeostasis: a matter of coordination. *Current opinion in neurobiology* 19:168–173.

Magee JC (1997) A Synaptically Controlled, Associative Signal for Hebbian Plasticity in Hippocampal Neurons. *Science* 275:209.

Magnusson D (1993) *Motor Development in Early and Later Childhood.*

Malenka RC (2003) The long-term potential of LTP. *Nat Rev Neurosci* 4:923–926.

Malenka RC, Bear MF (2004) LTP and LTD: an embarrassment of riches. *Neuron* 44:5–21.

Mann EO, Paulsen O (2007) Role of GABAergic inhibition in hippocampal network oscillations. *Trends in neurosciences* 30:343–349.

Marder E (2011) Variability, compensation, and modulation in neurons and circuits. *Proc Natl Acad Sci USA 108 Suppl* 3:15542–15548.

Marder E (2012) Neuromodulation of neuronal circuits: back to the future. *Neuron* 76:1–11.

Marder E, Goaillard J-M (2006) Variability, compensation and homeostasis in neuron and network function. *Nat Rev Neurosci* 7:563–574.

Marek KW, Ng N, Fetter R, Smolik S, Goodman CS, Davis GW (2000) A genetic analysis of synaptic development: pre- and postsynaptic dCBP control transmitter release at the Drosophila NMJ. *Neuron* 25:537–547.

Martin KC, Kandel ER (1996) Cell Adhesion Molecules, CREB, and the Formation of New Synaptic Connections. *Neuron* 17:567.

Masco D, Sahibzada N, Switzer R, Gale K (1999) Electroshock seizures protect against apoptotic hippocampal cell death induced by adrenalectomy. *Neuroscience* 91:1315–1319.

Mathern GW, Pretorius JK, Leite JP, Kornblum HI, Mendoza D, Lozada A, Bertram EH (1998) Hippocampal AMPA and NMDA mRNA levels and subunit immunoreactivity in human temporal lobe epilepsy patients and a rodent model of chronic mesial limbic epilepsy. *Epilepsy Res* 32:154–171.

Matthias K, Kirchhoff F, Seifert G, Hüttmann K, Matyash M, Kettenmann H, Steinhäuser C (2003) Segregated expression of AMPA-type glutamate receptors and glutamate transporters defines distinct astrocyte populations in the mouse hippocampus. *J Neurosci* 23:1750–1758.

Maximov A, Südhof TC, Bezprozvanny I (1999) Association of Neuronal Calcium Channels with Modular Adaptor Proteins. *J Biol Chem* 274:24453.

Mazzuferi M, Kumar G, Rospo C, Kaminski RM (2012) Rapid epileptogenesis in the mouse pilocarpine model: video-EEG, pharmacokinetic and histopathological characterization. *Exp Neurol* 238:156–167.

McAllister AK, Katz LC, Lo DC (1999) Neurotrophins and synaptic plasticity. *Annu Rev Neurosci* 22:295–318.

McKinney RA, Debanne D, Gähwiler BH, Thompson SM (1997) Lesion-induced axonal sprouting and hyperexcitability in the hippocampus in vitro: implications for the genesis of posttraumatic epilepsy. *Nat Med* 3:990–996.

Meador KJ (2007) The basic science of memory as it applies to epilepsy. *Epilepsia* 48 Suppl 9:23–25.

Meier JC, Henneberger C, Melnick I, Racca C, Harvey RJ, Heinemann U, Schmieden V, Grantyn R (2005) RNA editing produces glycine receptor alpha3(P185L), resulting in high agonist potency. *Nat Neurosci* 8:736–744.

Meldolesi J (1998) Calcium signalling. Oscillation, activation, expression. *Nature* 392.

Menniti FS, Chenard BL, Collins MB, Ducat MF, Elliott ML, Ewing FE, Huang JI, Kelly KA, Lazzaro JT, Pagnozzi MJ, Weeks JL, Welch WM, White WF (2000) Characterization of the binding site for a novel class of noncompetitive alpha-amino-3-hydroxy-5-methyl-4-isoxazolepropionic acid receptor antagonists. *Mol Pharmacol* 58:1310–1317.

Mermelstein PG, Bito H, Deisseroth K, Tsien RW (2000) Critical dependence of cAMP response element-binding protein phosphorylation on L-type calcium channels supports a selective response to EPSPs in preference to action potentials. *J Neurosci* 20:266–273.

Minor DL, Findeisen F (2010) Progress in the structural understanding of voltage-gated calcium channel (CaV) function and modulation. *Channels (Austin)* 4:459–474.

Molotkov D, Zobova S, Arcas JM, Khiroug L (2013) Calcium-induced outgrowth of astrocytic peripheral processes requires actin binding by Profilin-1. *Cell Calcium* 53:338–348.

Morrell F, deToledo-Morrell L (1999) From mirror focus to secondary epileptogenesis in man: an historical review. *Adv Neurol* 81:11–23.

Mulkey RM, Endo S, Shenolikar S, Malenka RC (1994) Involvement of a calcineurin/inhibitor-1 phosphatase cascade in hippocampal long-term depression. *Nature* 369:486–488.

Murthy VN, Schikorski T, Stevens CF, Zhu Y (2001) Inactivity Produces Increases in Neurotransmitter Release and Synapse Size. *Neuron* 32:673.

Nagelhus EA, Mathiisen TM, Ottersen OP (2004) Aquaporin-4 in the central nervous system: cellular and subcellular distribution and coexpression with KIR4.1. *Neuroscience* 129:905–913.

Naylor DE, Liu H, Niquet J, Wasterlain CG (2013) Rapid surface accumulation of NMDA receptors increases glutamatergic excitation during status epilepticus. *Neurobiol Dis* 54:225–238.

Naylor DE, Liu H, Wasterlain CG (2005) Trafficking of GABA(A) receptors, loss of inhibition, and a mechanism for pharmacoresistance in status epilepticus. *J Neurosci* 25:7724–7733.

Nerbonne JM, Gerber BR, Norris A, Burkhalter A (2008) Electrical remodelling maintains firing properties in cortical pyramidal neurons lackingKCND2-encoded A-type K+currents. *J Physiol (Lond)* 586:1565.

Newman EA (1993) Inward-rectifying potassium channels in retinal glial (Müller) cells. *J Neurosci* 13:3333–3345.

Nicoll RA, Schmitz D (2005) Synaptic plasticity at hippocampal mossy fibre synapses. *Nat Rev Neurosci* 6:863–876.

Nishida H, Okabe S (2007) Direct astrocytic contacts regulate local maturation of dendritic spines. *J Neurosci* 27:331–340.

Nusser Z, Cull-Candy S, Farrant M (1997) Differences in Synaptic GABAA Receptor Number Underlie Variation in GABA Mini Amplitude. *Neuron* 19:697.

Nusser Z, Hájos N, Somogyi P, Mody I (1998) Increased number of synaptic GABA(A) receptors underlies potentiation at hippocampal inhibitory synapses. *Nature* 395:172–177.

Oliet S, Panatier A, Piet R, Mothet J, Poulain D, Theodosis D (2008) *Neuron–glia interactions in the rat supraoptic nucleus.*

Ohno-Shosaku T, Maejima T, Kano M (2001) Endogenous Cannabinoids Mediate Retrograde Signals from Depolarized Postsynaptic Neurons to Presynaptic Terminals. *Neuron* 29:729.

O'Brien RJ, Kamboj S, Ehlers MD, Rosen KR, Fischbach GD, Huganir RL (1998) Activity-Dependent Modulation of Synaptic AMPA Receptor Accumulation. *Neuron* 21:1067.

Pak DTS, Sheng M (2003) Targeted protein degradation and synapse remodeling by an inducible protein kinase. *Science* 302:1368–1373.

Panatier A, Theodosis DT, Mothet J-P, Touquet B, Pollegioni L, Poulain DA, Oliet SHR (2006) Glia-derived D-serine controls NMDA receptor activity and synaptic memory. *Cell* 125:775–784.

Parrish JZ, Kim CC, Tang L, Bergquist S, Wang T, Derisi JL, Jan LY, Jan YN, Davis GW (2014) Krüppel mediates the selective rebalancing of ion channel expression. *Neuron* 82:537–544.

Patrushev I, Gavrilov N, Turlapov V, Semyanov A (2013) Subcellular location of astrocytic calcium stores favors extrasynaptic neuron-astrocyte communication. *Cell Calcium* 54:343–349.

Patsalos PN (2014) The clinical pharmacology profile of the new antiepileptic drug perampanel: A novel noncompetitive AMPA receptor antagonist. *Epilepsia* 56:12–27.

Peng Z, Zhang N, Wei W, Huang CS, Cetina Y, Otis TS, Houser CR (2013) A reorganized GABAergic circuit in a model of epilepsy: evidence from optogenetic labeling and stimulation of somatostatin interneurons. *J Neurosci* 33:14392–14405.

Penner MR, Pinaud R, Robertson HA (2001) Rapid kindling of the hippocampus protects against neural damage resulting from status epilepticus. *Neuroreport* 12:453.

Perea G, Navarrete M, Araque A (2009) Tripartite synapses: astrocytes process and control synaptic information. *Trends in neurosciences* 32:421–431.

Perez-Alvarez A, Navarrete M, Covelo A, Martin ED, Araque A (2014) Structural and functional plasticity of astrocyte processes and dendritic spine interactions. *J Neurosci* 34:12738–12744.

Pilgram GSK, Potikanond S, van der Plas MC, Fradkin LG, Noordermeer JN (2011) The RhoGAP crossveinless-c interacts with Dystrophin and is required for synaptic homeostasis at the Drosophila neuromuscular junction. *J Neurosci* 31:492–500.

Plomp JJ, van Kempen GT, Molenaar PC (1992) Adaptation of quantal content to decreased postsynaptic sensitivity at single endplates in alpha-bungarotoxin-treated rats. *J Physiol (Lond)* 458:487.

Pozo K, Goda Y (2010) Unraveling mechanisms of homeostatic synaptic plasticity. *Neuron* 66:337–351.

Pérez-Otaño I, Ehlers MD (2005) Homeostatic plasticity and NMDA receptor trafficking. *Trends in neurosciences* 28:229–238.

Racine RJ (1972) Modification of seizure activity by electrical stimulation. II. Motor seizure. *Electroencephalogr Clin Neurophysiol* 32:281–294.

Rogawski MA (2011) Revisiting AMPA receptors as an antiepileptic drug target. *Epilepsy Curr* 11:56–63.

Rollenhagen A, Sätzler K, Rodríguez EP, Jonas P, Frotscher M, Lübke JHR (2007) Structural determinants of transmission at large hippocampal mossy fiber synapses. *J Neurosci* 27:10434–10444.

Rongo C, Kaplan JM (2000) CaMKII regulates the density of central glutamatergic synapses in vivo. *Nature* 402:195–199.

Rugg-Gunn F (2014) Adverse effects and safety profile of perampanel: a review of pooled data. *Epilepsia* 55 Suppl 1:13–15.

Rusakov DA, Zheng K, Henneberger C (2011) Astrocytes as Regulators of Synaptic Function: A Quest for the Ca2+ Master Key. *Neuroscientist* 17:513.

Sahlender DA, Savtchouk I, Volterra A (2014) What do we know about gliotransmitter release from astrocytes? *Philos Trans R Soc Lond, B, Biol Sci* 369:20130592.

Saliba RS, Gu Z, Yan Z, Moss SJ (2009) Blocking L-type voltage-gated Ca2+ channels with dihydropyridines reduces gamma-aminobutyric acid type A receptor expression and synaptic inhibition. *J Biol Chem* 284:32544–32550.

Saliba RS, Michels G, Jacob TC, Pangalos MN, Moss SJ (2007) Activity-dependent ubiquitination of GABA(A) receptors regulates their accumulation at synaptic sites. *J Neurosci* 27:13341–13351.

Sanes JR, Lichtman JW (1999) Development of the vertebrate neuromuscular junction. *Annu Rev Neurosci* 22:389–442.

Schaub JR, Lu X, Doneske B, Shin Y-K, McNew JA (2006) Hemifusion arrest by complexin is relieved by Ca2+-synaptotagmin I. *Nat Struct Mol Biol* 13:748–750.

Schulman H, Heist K, Srinivasan M (1995) Decoding Ca2+ signals to the nucleus by multifunctional CaM kinase. *Prog Brain Res* 105:95–104.

Schuster CM, Davis GW, Fetter RD, Goodman CS (1996) Genetic Dissection of Structural and Functional Components of Synaptic Plasticity. I. Fasciclin II Controls Synaptic Stabilization and Growth. *Neuron* 17:641.

Serrano E, Kanner AM (2015) Recent treatment advances and novel therapeutic approaches in epilepsy. *F1000Prime Rep* 7:61.

Sharma AK, Reams RY, Jordan WH, Miller MA, Thacker HL, Snyder PW (2007) Mesial temporal lobe epilepsy: pathogenesis, induced rodent models and lesions. *Toxicol Pathol* 35:984–999.

Shi S (1999) Rapid Spine Delivery and Redistribution of AMPA Receptors After Synaptic NMDA Receptor Activation. *Science* 284:1811.

Srinivasan M, Edman CF, Schulman H (1994) Alternative splicing introduces a nuclear localization signal that targets multifunctional CaM kinase to the nucleus. *J Cell Biol* 4:387.

Stasheff S, Anderson W, Clark S, Wilson W (1989) NMDA antagonists differentiate epileptogenesis from seizure expression in an in vitro model. *Science* 245:648.

Stasheff SF, Bragdon AC, Wilson WA (1985) Induction of epileptiform activity in hippocampal slices by trains of electrical stimuli. *Brain Res* 344:296–302.

Steinhoff BJ (2015) The AMPA receptor antagonist perampanel in the adjunctive treatment of partial-onset seizures: clinical trial evidence and experience. *Ther Adv Neurol Disord* 8:137–147.

Steinhoff BJ, Hamer H, Trinka E, Schulze-Bonhage A, Bien C, Mayer T, Baumgartner C, Lerche H, Noachtar S (2014) A multicenter survey of clinical experiences with perampanel in real life in Germany and Austria. *Epilepsy Res* 108:986–988.

Stief F, Zuschratter W, Hartmann K, Schmitz D, Draguhn A (2007) Enhanced synaptic excitation-inhibition ratio in hippocampal interneurons of rats with temporal lobe epilepsy. *Eur J Neurosci* 25:519–528.

Striessnig J (2016) Voltage-gated calcium channels - from basic mechanisms to disease. *J Physiol (Lond)* 594:5817–5821.

Sugaya Y, Maru E, Kudo K, Shibasaki T, Kato N (2010) Levetiracetam suppresses development of spontaneous EEG seizures and aberrant neurogenesis following kainate-induced status epilepticus. *Brain Res* 1352:187–199.

Sun C, Sun J, Erisir A, Kapur J (2013) Loss of cholecystokinin-containing terminals in temporal lobe epilepsy. *Neurobiol Dis* 62:44–55.

Sutula TP (2004) Mechanisms of epilepsy progression: current theories and perspectives from neuroplasticity in adulthood and development. *Epilepsy Res* 60:161–171.

Suzuki A, Stern SA, Bozdagi O, Huntley GW, Walker RH, Magistretti PJ, Alberini CM (2011) Astrocyte-neuron lactate transport is required for long-term memory formation. *Cell* 144:810–823.

Swanwick CC, Murthy NR, Kapur J (2005) Activity-dependent scaling of GABAergic synapse strength is regulated by brain-derived neurotrophic factor. *Mol Cell Neurosci* 31:481–492.

Talley EM, Cribbs LL, Lee JH, Daud A, Perez-Reyes E, Bayliss DA (1999) Differential distribution of three members of a gene family encoding low voltage-activated (T-type) calcium channels. *J Neurosci* 19:1895–1911.

Tamaru Y, Nomura S, Mizuno N, Shigemoto R (2001) Distribution of metabotropic glutamate receptor mGluR3 in the mouse CNS: differential location relative to pre- and postsynaptic sites. *Neuroscience* 106:481–503.

Tanaka H, Shan W, Phillips GR, Arndt K, Bozdagi O, Shapiro L, Huntley GW, Benson DL, Colman DR (2000) Molecular modification of N-cadherin in response to synaptic activity. *Neuron* 25:93–107.

Tang J, Maximov A, Shin O-H, Dai H, Rizo J, Südhof TC (2006) A complexin/synaptotagmin 1 switch controls fast synaptic vesicle exocytosis. *Cell* 126:1175–1187.

Tedford HW, Zamponi GW (2006) Direct G protein modulation of Cav2 calcium channels. *Pharmacol Rev* 58:837–862.

Temporal S, Desai M, Khorkova O, Varghese G, Dai A, Schulz DJ, Golowasch J (2011) Neuromodulation independently determines correlated channel expression and conductance levels in motor neurons of the stomatogastric ganglion. *J Neurophysiol* 107:718–727.

Theodosis DT (2002) Oxytocin-secreting neurons: A physiological model of morphological neuronal and glial plasticity in the adult hypothalamus. *Front Neuroendocrinol* 23:101–135.

Thiagarajan TC, Lindskog M, Tsien RW (2005) Adaptation to synaptic inactivity in hippocampal neurons. *Neuron* 47:725–737.

Thomas KL, Laroche S, Errington ML, Bliss TVP, Hunt SP (1994) Spatial and temporal changes in signal transduction pathways during LTP. *Neuron* 13:737.

Thrane AS, Rappold PM, Fujita T, Torres A, Bekar LK, Takano T, Peng W, Wang F, Thrane VR, Enger R, Haj-Yasein NN, Skare Ø, Holen T, Klungland A, Ottersen OP, Nedergaard M, Nagelhus EA (2010) Critical role of aquaporin-4 (AQP4) in astrocytic Ca2+ signaling events elicited by cerebral edema. *Proc Natl Acad Sci USA* 108:846–851.

Turrigiano G (2011) Too many cooks? Intrinsic and synaptic homeostatic mechanisms in cortical circuit refinement. *Annu Rev Neurosci* 34:89–103.

Turrigiano G (2012) Homeostatic synaptic plasticity: local and global mechanisms for stabilizing neuronal function. *Cold Spring Harb Perspect Biol* 4:a005736.

Turrigiano GG (1999) Homeostatic plasticity in neuronal networks: the more things change, the more they stay the same. *Trends in neurosciences* 22:221.

Turrigiano GG (2000) AMPA Receptors Unbound. *Neuron* 26:5.

Turrigiano GG (2008) The self-tuning neuron: synaptic scaling of excitatory synapses. *Cell* 135:422–435.

Turrigiano GG, Leslie KR, Desai NS, Rutherford LC, Nelson SB (1998) Activity-dependent scaling of quantal amplitude in neocortical neurons. *Nature* 391:892–896.

Turrigiano GG, Nelson SB (2004) Homeostatic plasticity in the developing nervous system. *Nat Rev Neurosci* 5:97–107.

Tzingounis AV, Wadiche JI (2007) Glutamate transporters: confining runaway excitation by shaping synaptic transmission. *Nat Rev Neurosci* 8:935–947.

Vertes RP, Hoover WB, Do Valle AC, Sherman A, Rodriguez JJ (2006) Efferent projections of reuniens and rhomboid nuclei of the thalamus in the rat. *J Comp Neurol* 499:768–796.

Vithlani M, Terunuma M, Moss SJ (2011) The dynamic modulation of GABA(A) receptor trafficking and its role in regulating the plasticity of inhibitory synapses. *Physiol Rev* 91:1009–1022.

Volterra A, Liaudet N, Savtchouk I (2014) Astrocyte Ca^{2+} signalling: an unexpected complexity. *Nat Rev Neurosci* 15:327–335.

Wan J (1999) Activity-Induced Potentiation of Developing Neuromuscular Synapses. *Science* 285:1725.

Wang H, Kunkel DD, Schwartzkroin PA, Tempel BL (1994) Localization of Kv1.1 and Kv1.2, two K channel proteins, to synaptic terminals, somata, and dendrites in the mouse brain. *J Neurosci* 14:4588–4599.

Watt AJ, Desai NS (2010) Homeostatic Plasticity and STDP: Keeping a Neuron's Cool in a Fluctuating World. *Front Synaptic Neurosci* 2:5.

Weisskopf MG, Bauer EP, LeDoux JE (1999) L-type voltage-gated calcium channels mediate NMDA-independent associative long-term potentiation at thalamic input synapses to the amygdala. *J Neurosci* 19:10512–10519.

Wenner P (2011) Mechanisms of GABAergic homeostatic plasticity. *Neural Plast* 2011:489470.

Wenzel HJ, Vacher H, Clark E, Trimmer JS, Lee AL, Sapolsky RM, Tempel BL, Schwartzkroin PA (2007) Structural consequences of Kcna1 gene deletion and transfer in the mouse hippocampus. *Epilepsia* 48:2023–2046.

Wilson RI, Kunos G, Nicoll RA (2001) Presynaptic Specificity of Endocannabinoid Signaling in the Hippocampus. *Neuron* 31:453.

Winkelmann A, Maggio N, Eller J, Caliskan G, Semtner M, Häussler U, Jüttner R, Dugladze T, Smolinsky B, Kowalczyk S, Chronowska E, Schwarz G, Rathjen FG, Rechavi G, Haas CA, Kulik A, Gloveli T, Heinemann U, Meier JC (2014) Changes in neural network homeostasis trigger neuropsychiatric symptoms. *J Clin Invest* 124:696–711.

Witcher MR, Kirov SA, Harris KM (2006) Plasticity of perisynaptic astroglia during synaptogenesis in the mature rat hippocampus. *Glia* 55:13.

Wolfart J, Laker D (2015) Homeostasis or channelopathy? Acquired cell type-specific ion channel changes in temporal lobe epilepsy and their antiepileptic potential. *Front Physiol* 6:168.

Wouterlood FG, Saldana E, Witter MP (1990) Projection from the nucleus reuniens thalami to the hippocampal region: light and electron microscopic tracing study in the rat with the anterograde tracer Phaseolus vulgaris-leucoagglutinin. *J Comp Neurol* 296:179–203.

Wykes RC, Heeroma JH, Mantoan L, Zheng K, MacDonald DC, Deisseroth K, Hashemi KS, Walker MC, Schorge S, Kullmann DM (2012) Optogenetic and potassium channel gene therapy in a rodent model of focal neocortical epilepsy. *Sci Transl Med* 4:161ra152.

Yang Y, Gozen O, Watkins A, Lorenzini I, Lepore A, Gao Y, Vidensky S, Brennan J, Poulsen D, Park JW, Jeon NL, Robinson MB, Rothstein JD (2009) Presynaptic regulation of astroglial excitatory neurotransmitter transporter GLT1. *Neuron* 61:880–894.

Yelshanskaya MV, Singh AK, Sampson JM, Narangoda C, Kurnikova M, Sobolevsky AI (2016) Structural Bases of Noncompetitive Inhibition of AMPA-Subtype Ionotropic Glutamate Receptors by Antiepileptic Drugs. *Neuron* 91:1305–1315.

Zamponi GW, Currie KPM (2012) Regulation of Ca(V)2 calcium channels by G protein coupled receptors. *Biochim Biophys Acta* 1828:1629–1643.

Zeilhofer HU, Studler B, Arabadzisz D, Schweizer C, Ahmadi S, Layh B, Bösl MR, Fritschy J-M (2004) Glycinergic neurons expressing enhanced green fluorescent protein in bacterial artificial chromosome transgenic mice. *J Comp Neurol* 482:123–141.

Zhao C, Dreosti E, Lagnado L (2011) Homeostatic synaptic plasticity through changes in presynaptic calcium influx. *J Neurosci* 31:7492–7496.

In: Synaptic Plasticity
Editor: Sraboni Chaudhury

ISBN: 978-1-53613-606-7

Chapter 9

NEURONAL ACTIVITY AND SYNAPTIC PLASTICITY UNDERLYING HIPPOCAMPAL MEMORY ENGRAM

Anupratap Tomar[1,2,*], PhD and Lalitha Krishnan[3]

[1] School of Physiology, Pharmacology and Neuroscience, University of Bristol, UK

[2] Laboratory for Circuit and Behavioral Physiology, RIKEN Brain Science Institute, Japan

[3]National Centre for Biological Sciences, Bangalore, India

ABSTRACT

Activity-dependent synaptic plasticity is considered to be the physiological mechanism underlying learning and memory. Since the discovery of long-term potentiation (LTP) in 1973 by Bliss and Lomo, the popularity of synaptic plasticity has grown immensely. By October 2016, more than 51,000 research articles and reviews were indexed

---

* Corresponding Author: anupratap.tomar@bristol.ac.uk.

by PubMed. The recent emergence of optogenetics and chemogenetics has enabled researchers to generate neuronal activity patterns with precise temporal and spatial control in awake animals and brain slices. The relative ease of combining these tools with traditional electrophysiology will further expand the scope of synaptic plasticity research in freely behaving animals while they engage in cognition and memory tasks. This chapter aims to provide an overview of recent advances in the field of synaptic plasticity and learning and memory with a focus on cognitive impairments caused by stress-induced maladaptive synaptic plasticity. We summarize the latest memory engram data that claims to specifically manipulate the activity of hippocampal memory cells by employing optogenetics and DREADD chemogenetics. We end by discussing the potential consequences of long term usage of circuit genetic tools on neuronal network dynamics.

Keywords: synaptic plasticity, memory, cognition, optogenetics, engram, DREADDs, stress

1. Introduction

Where is information stored in the brain and how is it recalled at appropriate times are questions that have fascinated researchers for centuries. By now it is well established that there exist various memory modules in mammalian brains such that different types of memories engage different brain regions. Human fMRI studies have established that the hippocampus, a medial temporal lobe structure, is activated not only during declarative learning, such as about names, facts and spatial locations, but also during their recall (Carr et al., 2010). These observations clearly suggest that hippocampal neuronal networks support memory encoding and its recall. But where in the hippocampus is the memory encoded? Does a specific hippocampal cell represent a specific memory event, or are a set of neurons involved in encoding and recall of memory representations? Previous research suggests that the latter possibility may be true. Thus, memory researchers have focused on changes in the synapses, the communication point between neurons, after learning as the possible locus of memory representations. It is well established that synapses are malleable and that they undergo molecular and structural modifications in an activity-dependent manner i.e., after learning as well as after delivery of certain electrical activity patterns (Morris, 1989; Takeuchi et al., 2014). Synaptic plasticity caused in an activity-dependent manner can thus modify neuronal communication and activity after learning and is hypothesized to be the mechanism underlying learning and memory.

This chapter aims to synthesize the latest research on rodent declarative memory and discusses various studies that have tested the necessity and sufficiency of hippocampal cell activity and synaptic plasticity in formation and recall of memory. We highlight recent advances in localizing and manipulation of the physical trace of a memory in hippocampus-mediated learning tasks. Throughout this chapter, we will interchangeably use terms such as engram, engram cells and memory trace to describe learning-associated putative memory cells in the mouse hippocampus. Because of space constraints, only a few selected studies that are either considered very impactful in the field or were the first of their kind are discussed.

We start with a brief historical run-through of the role of the hippocampus in synaptic plasticity and learning and memory research. Then we provide a brief introduction to hippocampal neuronal subnetworks and connectivity. Using the specific case of hippocampal place cells, we discuss the evidence of modification of synaptic plasticity after spatial learning. Next, we focus on immediate early gene (IEG) imaging highlighting c-Fos expression that occurs within a few minutes after neuronal activity thus providing a method to identify recently active cells. We describe c-Fos-tTA-reporter technology that can tag putative memory engram cells in the hippocampus and elsewhere. Then we discuss the recent advent in bidirectional cell-activity manipulation technology such as optogenetics and DREADD chemogenetics. These tools in combination with C-Fos-tTA technology have been employed to study the causal link between engram cell activity, synaptic plasticity and memory recall. We conclude the review by summarizing various sections and then briefly describe the different directions that engram manipulation research is moving in.

2. Brief Historical Overview of Memory and Synaptic Plasticity Research

Employing the Golgi staining method, Ramon Y Cajal first described in his neuron doctrine, that the brain is made up of billions of discrete cells called neurons. Cajal also speculated that memories could be stored as new connections between neurons (Cajal, 1894; Mayford et al., 2012). This hypothesis is similar to that proposed by German psychologist Richard Semon, who stated that psychological states like memory leave a physiological mark (some form of alteration in neurons) in the brain. He coined the term 'engram' to define the trace of a memory (Schacter et al., 1978; Semon, 1921). In the 1920s, Karl Lashley set out to search for the location of these engrams using brain lesion studies in rats. He would perform graded brain lesions and subsequently test a rat's performance on memory tasks in mazes. Eventually, Lashley concluded that memory traces or engrams are not stored in a specific part of the brain but are distributed throughout (Lashley, 1950). However, less than a decade later, the curious case of amnesiac patient Henry Gustav Molaisen (H.M.) partially challenged this claim. H.M. could not form new declarative memories after the surgical removal of his medial temporal lobe, that included the densely packed seahorse-shaped brain region called the hippocampus (Scoville and Milner, 1957). H.M.'s condition suggested that the hippocampus played a crucial role in encoding and consolidation of declarative memory and brought the hippocampus into the focus of cognitive neuroscience research.

The information acquired during the learning process is thought to be stored by either modification of synaptic strength or a change in the number of synapses or both. The year 1949 saw the formulation of the first major theory on information storage and recall by Canadian psychologist Donald Hebb. In his book *The organization of behaviour:*

A psychological theory, Hebb postulated that during memory acquisition physically connected neurons (via synapses) form functional cell assemblies when they are simultaneously activated by an external stimulus (Hebb, 1949). Hebb hypothesized that synapses which had become part of a cell assembly should experience an increase in synaptic strength. Subsequently, when a similar activity pattern activated a subset of these neurons, the whole cell assembly could eventually be activated in which case it is likely the subject would recall the event.

By the late 1950s, the hippocampus had been established as the crucial region for memory encoding and researchers were now searching for a cellular correlate underlying learning and memory in experimental animals. Then in the 1970s, Lomo and Bliss found evidence of electrical activity-dependent synaptic strengthening in the hippocampus (Bliss and Lomo, 1973). To mimic the transmission of information into the hippocampus, Bliss and Lomo had electrically stimulated hippocampal afferents with high frequency activity patterns (theta activity). Because the resultant increase in synaptic strength was persistent, this phenomenon was termed as Long-Term-Potentiation (LTP).

Since the discovery of LTP, the hippocampus has become the major focus of research on cellular and molecular mechanisms underlying synaptic plasticity. It is now well established that activity-dependent synaptic plasticity is a prominent feature of hippocampal synapses. Both sides of the synapse undergo modification. The quantity of neurotransmitter released changes in the pre-synaptic side, while post-synaptic alterations include alterations in the number of ionic receptors and their conductance. Hippocampal LTP occurs concomitantly with learning and memory (Whitlock et al., 2006). While LTP is caused by higher bursts of electrical activity, Long-term depression (LTD) is cause by low frequency (1 Hz) electrical stimulation of hippocampal inputs. The electrical stimulation of hippocampal inputs at low frequency (1 Hz), leads to long-term weakening of the post-synaptic response (Dudek and Bear, 1992; Dunwiddie and Lynch, 1978), and is also hypothesized to play a crucial role in memory.

3. Neuroanatomy of the Hippocampal System

The hippocampus includes a compactly packed cell layer that is called Cornu Ammonis (CA) and is divided into three subfields (CA1-CA3). The input to CA3 region arrives from the neighbouring dentate gyrus (DG) via mossy fibers (MF). CA3 cells synapse on to CA1 cells via schaffer collaterals (SC). Together CA subregions (CA1, CA2, CA3) and DG form the hippocampus or hippocampal formation whose synaptic plasticity and function in learning and memory we discuss in this chapter. The hippocampus has been in the focus of information processing research due to its largely unidirectional connectivity (DG→CA3→CA1). The rodent hippocampus has been estimated to have ~120,000-250,000 cells each in areas CA1 and CA3, while the DG has been estimated to house ~1,000,000 neurons. Based on synaptic plasticity data from transgenic mice, lesion studies and computational modeling work, the DG has been assigned a role in pattern separation (Leutgeb et al., 2007; McHugh and Tonegawa, 2007). CA3 is said to be involved in pattern completion, temporary memory storage and recall (Marr, 1971; McNaughton and Morris, 1987; Nakazawa et al., 2004). Since CA1, the output region of the hippocampus, receives inputs from CA3 as well as the cortex (entorhinal cortex, EC), it can compare the current information arriving from EC and processed information from CA3 and is said to work as a comparator and novelty detector (O'Reilly and Rudy, 2001; Vinogradova, 2001). A lot of recent work has been done to describe the role of DG neurogenesis in encoding of new information (Nakashiba et al., 2012) as well as in forgetting (Epp et al., 2016; Frankland et al., 2013). Hence, DG appears to be uniquely placed to influence hippocampus-mediated learning and we will discuss about DG engram cell manipulation studies in more detail in later sections.

4. Hippocampal Pyramidal Cells Undergo Synaptic Strength Modificataions During Encoding Of Space and Context

Neurons produce small electrical pulses (spikes) called action-potentials. This electrical chit-chat between neurons can be recorded in live animals, by placing microelectrodes close to neuronal cell bodies. This in vivo electrophysiology technique was employed by John' O Keefe and Dostrovsky, who noticed a curious phenomenon; that hippocampal cells displayed enhanced spiking only when rats visited a specific location in their surroundings (O'Keefe and Dostrovsky, 1971). They termed these hippocampal cells as 'place cells'. Later, O'Keefe and Lynn Nadel proposed that place cells could represent different areas in the animal's surroundings thereby forming an internal map which they called the cognitive map (O'Keefe and Nadel, 1978). It is now established that approximately 30-50% of pyramidal cells display characteristic place cell activity in any given environment (Thompson and Best, 1990). The evidence supporting the hypothesis that the hippocampus transmits an ensemble code for location includes the observation that activity of ~80-150 hippocampal place cells can successfully predict an animal's location in its surroundings (Wilson and McNaughton, 1993). It was observed that during exposure to a novel environment the prediction of the animal's location (place code) was poor but improved rapidly with experience. Further, cells that fired together during spatial encoding (i.e., cells encoding common locations) displayed an increased tendency to fire together during a subsequent rest period post-exploration, suggesting occurrence of synaptic potentiation during learning (Wilson and McNaughton, 1994).

Moser and colleagues reported experience-induced enhanced synaptic potentials in the DG-perforant pathway in rodents, further linking spatial learning and synaptic strengthening in the hippocampus (Moser et al., 1994). Additionally, transgenic mice with impaired NMDA receptor-mediated synaptic plasticity not only displayed less-specific CA1 place cell activity but also showed extremely low co-firing for cell pairs with overlapping place fields suggesting poor spatial information transfer to downstream brain regions (McHugh et al., 1996). No surprise that these mice displayed poor performance on hippocampus-mediated spatial learning tasks. Another experiment that correlated place cell activity to Hebbian synaptic plasticity involved first identifying CA1 and CA3 place cell pairs with overlapping place fields in freely behaving mice and then replaying this activity pattern in vitro (Isaac et al., 2009). This procedure led to a robust and rapid induction of NMDAR-dependent LTP in CA1 pyramidal cells in the presence of carbachol which provided cholinergic tone to hippocampal networks.

We tend to consider our surroundings as the same even if a few changes are made. However, if these changes are too many, the environment may be perceived as distinct from the previous one. How are different surroundings and contexts represented in the hippocampus? Two studies in year 2005 addressed this issue by assessing place cell remapping in gradually changing environments (Leutgeb et al., 2005; Wills et al., 2005). Place cells may encode two different environments by changing their preferred location of spiking (place remapping) or by modifying the level of spiking, either enhancing or reducing it (rate remapping). At the population level, these two types of remapping give rise to a phenomenon called global remapping, which is measured by population vector analysis. While Wills and colleagues (2005) reported a gestalt shift of population vectors in gradually changing surroundings, Leutgeb et al., (2005) noticed a gradual shift in place cell population activity. Regardless of this discrepancy in their results, population vector analysis provided a good handle on assessing a rodent's ability to discriminate between environments. Using population vector analysis, it was reported that chronically stressed mice showed reduced global remapping in similar contexts (Tomar et al., 2015) suggesting that stress impairs perception of surroundings, a symptom of post-traumatic-stress-disorders (PTSD).

5. Hippocampal Cells That Engage in Information Coding Are Identified By Using Immediate Early Gene (IEG) Imaging Technique

Specific memories or sensory representations should be sparsely encoded by a small fraction of neurons. By identifying these cells, Semon's memory trace, engram or Hebb's cell assemblies can then be visualized and manipulated. IEGs are rapidly and reliably expressed in response to robust cell activity (Hess et al., 1995). IEGs have been used to identify cellular activity and synaptic plasticity in hippocampus-mediated tasks such as radial-arm maze navigation, exposure to novel environment and contextual fear conditioning (Guzowski et al., 1999; Hall et al., 2001; Vann et al., 2000). While transcription regulatory IEGs such as c-Fos, c-Jun and Zif268 are markers of post-synaptic neuronal activity, effector genes such as Arc and Homer are the better markers of synaptic plasticity as they are involved in transcription of new synaptic proteins. Employing Arc it has been shown that coding is very sparse in DG as ~2-4% of cells display Arc expression (Chawla et al., 2005), however a large fraction (~40-45%) of CA1 cells display Arc activity (Guzowski et al., 1999). Interestingly, approximately the same fraction of cells is activated either during novel context encoding or during contextual fear learning suggesting valence of information coding does not play a crucial role in hippocampal information coding. The functional dynamics of Arc and c-Fos are similar (Guzowski et al., 2001) and c-Fos gene expression has been shown to affect hippocampal synaptic plasticity and memory recall in hippocampus-mediated fear conditioning (Fleischmann et al., 2003; Zhang et al., 2002). Due to the above-mentioned reasons, c-Fos emerged as the prominent IEG around which technology (described below) was built to identify and manipulate the activity of hippocampal engram cells.

6. Putative Memory Trace Is Visualized by c-Fos-tTA-Reporter Technology

As described earlier, the c-Fos imaging method can identify cells that were likely active during learning. The same set of cells are likely to be reactivated during memory recall. Since IEG expression remains heightened for less than an hour and degrades afterwards, IEG imaging method places a time constraint on when memory recall can be tested after learning. A simple solution to this problem involved using a transgenic method where the promoter region of c-Fos was linked to bacterial gene tauLacZ (LAC) which produced a reporter protein, β-galactosidase (Wilson et al., 2002). This c-Fos-reporter protein-dependent tagging method has proved effective in identifying putative memory cells associated with a specific percept or cognitive experience (Barth et al., 2004; Wang et al., 2006).

A major challenge that still remained was to prevent disappearance of the memory cells in a sea of tagged non-memory cells that were activated by unrelated behaviour in the interim between learning and recall. Generation of a c-Fos-tTA/TetO-tauLacZ transgenic mouse involving an inducible and persistent expression of reporter protein addressed this issue (Reijmers et al., 2007). Through transgenic approach, c-Fos promoter was linked to transcription of the tetracycline transactivator (tTA) gene. tTA protein binds to another transgene with a tet-O promoter and induces expression of downstream reporter LAC gene in an activity-dependent manner. One may wonder, how this tagging is restricted to the learning phase. The answer lies in the fact that tTA protein was rendered inactive in the presence of Doxycycline (Dox), and hence it could not bind to the tet-O promoter. These mice were regularly kept on a Dox diet (ON-DOX) and LAC was not expressed. Now just before learning, the Dox was removed from the diet (OFF-DOX) and this allowed the tTA protein to bind to the tet-O promoter and LAC was expressed. This procedure ensured that only those neurons that were active during learning got tagged with reporter protein β-galactosidase. Immediately after learning, these mice were put back on Dox diet (ON-DOX). This Dox-dependent c-Fos-tTA expression thus ensured exclusive tagging of putative memory cells. Now to solve the issue of persistent tagging, a mutated version of the tTA gene was attached to the tet-O promoter which expressed an altered version of tTA protein (say tTA*) that was insensitive to Dox and was constitutively active. Thus learning-associated expression of c-Fos promoter, in absence of Dox, would first bind to tet-O promoter and allowed expression of tTA* which in turn kept a continuous turnover of LAC and hence persistent tagging of putative memory cells was achieved even when back ON-DOX.

Using c-Fos-tTA/TetO-tauLacZ transgenic mice, Reijmers and colleagues could demonstrate that ~12% of basolateral amygdala (BLA) neurons were tagged with β-galactosidase during amygdala-mediated fear learning. Further, using another IEG Zif268, these researchers demonstrated that almost an equal fraction of BLA cells became reactivated during fear memory recall (ZIF^{+ve}). Hippocampal studies employing c-Fos-tTA/TetO-ChR2-EYFP transgenic mice reported that, like Arc results (Chawla et al., 2005), ~2-6% of DG cells displayed c-Fos expression if exposed to either a novel context or after fear-conditioning (Liu et al., 2012). Further, using c-Fos-tTA/TetO-ChR2-mcherry ~1.5% of DG cells and ~ 45% of CA1 cells displayed c-Fos activity when re-exposed to the same context (Ramirez et al., 2013). Overall, the above data strongly suggests that c-fos-tTA-reporter technology not only persistently tags learning-associated engram cells but many of these cells are reactivated during memory recall.

7. Optogenetics Technology Combines Fiber-Optics and Genetic Modification Tools to Bidirectionally Regulate Brain Cell Activity

As mentioned above, by 2007, there existed c-Fos-tTA-reporter technology that could persistently tag cells activated during learning. However, to be able to prove that activity of these cells drives memory recall, researchers needed a method to artificially activate these cells. Fortunately, two years prior to the advent of c-fos-tTA-reporter technology, optogenetics, a method that allows control of neuronal activity in genetically modified rodents by shining light on neurons had already been developed (Boyden et al., 2005). In this study through a transgenic approach, researchers introduced a gene that could produce and integrate a light-sensitive protein called channelrhodopsin-2 (ChR2, a cation selective ion channel) on neuronal membranes. Brain slices were prepared, and spiking activity was recorded while short pulses of blue light (473 nm) were delivered through laser. Light delivery led to an influx of cations into cells thus depolarizing these neurons which caused cell spiking. ChR2-expressing neurons displayed enhanced spiking only during laser pulse delivery thus providing clear evidence that neuronal activity can be regulated in a precise spatiotemporal manner. These experiments have been replicated in live animals and optogenetic stimulation augments hippocampal neuronal network activity (Takata et al., 2015). This includes heightened multiunit activity (MUA) and power of local field potentials (LFP) at theta, gamma, and ultra-fast frequency bands in CA1 neuronal networks. Optogenetic methods have also been employed to suppress neuronal activity by introducing light-sensitive proteins such as archeorhodopsin (Arch and ArchT, outward driven proton pumps) and halorhodopsin (NphR, an inward driven Cl- pump) that hyperpolarize neurons (Chow et al., 2012).

By changing the frequency of laser-pulse stimulation, the frequency of neuronal network activity can be manipulated. This ability to generate specific activity patterns in presynaptic neurons through ChR2 optogenetics, provides an ideal tool to study synaptic plasticity and memory in awake animals which was earlier difficult to achieve through crude electrical stimulation methods. A 2014 study investigated the causal link between synaptic plasticity and fear memory by optogenetic methods (Nabavi et al., 2014). Laser stimulation was delivered to BLA neuronal networks while the rats received mild foot-shock. Through this classical conditioning method, neuronal activity became associated with foot-shock such that a mere delivery of light would induce freezing behaviour, a readout of fear due to impending foot-shock. Brain slices prepared from these rats displayed enhanced NMDA/AMPA ratio in BLA, suggesting occurrence of Hebbian synaptic plasticity. In another experiment, a day after animals received laser-foot-shock pairing, they also received an optical LTD protocol (2 ms pulses of light, at 1 Hz). Incredibly, next day these rats did not show light-induced freezing indicating that fear memory was suppressed. The above data clearly demonstrates that optogenetics technology not only bidirectionally regulates neuronal activity but can be used to probe the link between synaptic plasticity and memory.

8. Chemogenetics Technology Combines Modified Molecules and Genetic Modification Tools to Influence Brain Cell Activity

Another tool that bidirectionally regulates neuronal activity is called Designer Receptors Exclusively Activated by Designer Drugs (DREADDs). DREADDs are genetically modified G-protein coupled receptors (GPCR) which are made insensitive to their endogenous ligand and are exclusively activated by inert ligands such as clozapine-N-Oxide (CNO). Based on the type of G-protein that these DREADDs interact with, they can either enhance (via Gq or Gs signaling) or suppress (via Gi signaling) neuronal activity. In the year 2007, a study expressed modified Gi-coupled human M4 muscarinic acetylcholine DREADD receptors (hM4Di) in hippocampal neurons and induced hyperpolarization thus suppressed neuronal activity upon CNO delivery (Armbruster et al., 2007). The same research group later created hM3Dq transgenic mice which expressed Gq coupled human M3 muscarinic receptors (hM3Dq) in the hippocampus and other cortical regions. CNO injection (i.p.) led to enhanced neuronal activity and these mice also showed increased locomotion and limbic seizures (Alexander et al., 2009). Further, DREADDs-CNO system also affects synaptic plasticity. A recent study reported that hippocampal slices prepared from mice injected with hM3Dq

DREADDs displayed LTP as enhanced slope and amplitude of field excitatory post-synaptic potentials (fEPSP) which was observed in area CA1 after 20 min of CNO application (López et al., 2016). This study also reported that hM4Di DREADDs caused impaired hippocampal LTP after CNO application. Thus DREADDs-CNO system has been shown to bidirectionally regulate hippocampal neuronal activity as well as hippocampal synaptic plasticity.

9. Artificial Activation of Putative Memory "Engram" Cells Induces Memory Recall

Using c-Fos-tTA-reporter technology, in year 2012, two studies claimed that artificial activation of memory engram cells is sufficient for hippocampus-mediated context fear memory recall (Garner et al., 2012; Liu et al., 2012). Garner and colleagues (2012) used hM3Dq-CNO system for cell activity manipulation. During OFF-Dox, c-Fos-tTA/TetO-hM3Dq transgenic mice were exposed to a novel context (Context-A) which led to tagging of active cells with hM3Dq receptors and mCherry reporter protein. Next day, during ON-DOX, Context-A engram cell activity induced by CNO injection was paired with foot-shock (fear conditioning) in another novel context (Context-B). This conditioning procedure led to an association between Context-A engram cell activity and foot-shock such that future reactivation of Context-A engram, by CNO, should have led to freezing in anticipation of the impending foot-shock. Surprisingly, these mice displayed freezing response only when they received CNO injection in Context-B but not in Context-A. Also, they did not show freezing in Context-B in absence of CNO. Memory recall had to specifically satisfy AND condition i.e., simultaneous activation of Context-A engram by CNO and activation of Context-B engram by exposure to Context-B. Thus, Garner and colleagues concluded that fear memory in these mice was represented by a hybrid memory trace such that Context-A engram was embedded in Context-B engram. This hybrid memory trace may represent autobiographical memories where very often one incident reminds us of another related incident and so on.

Liu and colleagues (2012) specifically focused on testing the role of hippocampal engrams (DG and CA1) in context fear memory recall by employing an optogenetics approach. During OFF-DOX, c-Fos-tTA/TetO-ChR2-EYFP transgenic mice were exposed to a novel context (Context-B) and received foot-shock (fear conditioning) in the same context. This conditioning procedure led to tagging of cells active in Context-B with ChR2-EYFP and also associated the activity of these cells with foot-shock. Now during subsequent ON-DOX period, reactivation of Context-B engram, by laser (blue light) stimulation, lead to freezing in anticipation of the impending foot-shock in a previously familiar neutral context (Context-A). This study demonstrated the causal link between engram cell activity and memory recall such that artificial reactivation of fear-associated engram cells led to context fear recall.

In a subsequent study, the same group examined if c-Fos-tTA/TetO-ChR2-mCherry mice could be tricked into fearing a previously familiar but neutral context without artificially reactivating the fear-associated engram cells during memory recall (Ramirez et al., 2013). During OFF-DOX, mice were exposed to a novel context (Context-A) and cells involved in encoding of Context-A got tagged with ChR2-mCherry. Next day during ON-DOX, Context-A engram cell reactivation, by blue-light stimulation, was paired with foot-shock (fear conditioning) in another novel context (Context-B). As expected, subsequent exposure to Context-A, naturally reactivated Context-A engram cells (without blue light stimulation) which predicted the impending foot-shock and hence mice now froze in Context-A. Since mice displayed freezing response in a context where they had never received foot-shock, and nor was there artificial reactivation of fear-associated engram, this experiment was interpreted as the one that implanted a false memory in mice.

Recall that Garner and colleagues in their hybrid memory experiment failed to observe any fear in an original neutral context where the engram-labelling had occurred (Garner et al., 2012). How can this discrepancy between synthetic memory experiments by Garner and colleagues and

Rameriez at al. (Ramirez et al., 2013), be explained? Unlike Ramirez and colleagues who optically activated Context-A DG engram for very short durations with foot-shock, CNO injection in c-Fos-tTA/TetO-hM3Dq transgenic mice in Garner et al's study underwent a sustained and prolonged reactivation of Context-A engram before and after foot-shock delivery in Context-B. Further, while the false memory study involved labeling and reactivation of DG engram cells, expression of hM3Dq receptors was brain wide. Thus, mismatch of temporal and spatial specificity in engram cell manipulation during fear conditioning may have something to do with conflicting results obtained by these two studies.

10. Artificial Silencing of Putative Memory "Engram" Cell Activity Impairs Memory Recall

In the year 2014, three studies tested the necessity of hippocampal engram cell activity in context fear recall (Denny et al., 2014; Tanaka et al., 2014; Zhu et al., 2014). Denny and colleagues (2014) used an optogenetic approach and labeled engram cells in area CA3 or DG with Arch-EYFP during foot-shock delivery in Context-A (Conditioning). 2 weeks later, when tested for fear recall by placing these mice into Context-A, while simultaneously suppressing engram cell activity through the usage of yellow light (593 nm), these mice expressed reduced freezing response suggesting muted fear recall. Similarly, Tanaka and colleagues (2014) used c-Fos-tTA/TetO-ArchT-H2B-GFP transgenic mice and during OFF-DOX, labeled CA1 engram cells with Fos-H2B-GFP-ArchT in Context-A during fear conditioning. Later during fear recall, when the activity of engram cells was suppressed by yellow laser, these mice failed to express fear recall as they displayed much lesser freezing than control mice expressing Fos-H2B-GFP but no ArchT. A subsequent study employed hM4Di DREADDs to suppress the activity of ventral hippocampal CA1 cells within 6 hours after conditioning (during memory

consolidation period) and revealed that fear memory was impaired (Zhu et al., 2014).

11. Structural and Synaptic Plasticity Differences between c-Fos Expressing "Engram" Cells and c-Fos Negative Cells

While majority of engram studies focus on regulating the activity of c-Fos expressing engram cells, very few studies have probed the effects of a learning episode on structural and synaptic modifications occurring in engram cells. For example, CA1 engram cells of rodents that had undergone hippocampus-mediated contextual fear conditioning had fewer dendritic spines than that of home cage control group animals (Sanders et al., 2012). Surprisingly, another study that focused on DG engram cells found that c-Fos expressing neurons displayed a greater number of dendritic spines (Ryan et al., 2015). This study however did not examine spine density in area CA1. Further, Ryan and colleagues also found that fear conditioning affects hippocampal synaptic plasticity as DG engram cells also showed a greater number of AMPA receptor insertions as well as higher spontaneous EPSC frequencies and amplitudes. Furthermore, this synaptic plasticity was protein synthesis-dependent as anisomysin application led to blocking of the above-mentioned changes in c-Fos engram cells. BLA engram cells are also reported to have enhanced EPSC frequencies suggesting presynaptic potentiation onto engram cells after learning (Nonaka et al., 2014). These data suggest that learning-associated c-Fos expressing cells indeed undergo structural and synaptic plasticity and disparate neuronal networks show differences in magnitude and direction of learning-induced plasticity.

CONCLUSION

Reactivation of learning-associated activity patterns have been hypothesized to induce memory recall (Hebb, 1949; Martin and Morris, 2002). Hebb postulated that during learning, a pattern of neural activity induced by an event causes synapses to undergo synaptic strengthening and that reactivation of these learning-associated synapses should occur during memory recall. In this chapter, we discussed the evidence that artificial activation of learning-associated neurons (engram cells) indeed induces context-specific freezing behaviour, a readout of hippocampus-mediated fear memory recall. One may wonder, and rightly so, if artificial activation of hippocampal memory engram cells induces the same qualitative experience as was felt during learning. The answer most likely is going to be no but there is no doubt that some aspects of the event are indeed recalled. By comparing synthetic memory experiments (Garner et al., 2012) using DREADD-CNO approach to false memory experiments and employing ChR2-optogenetics (Ramirez et al., 2013), we pointed out that long-duration activation of memory trace provided different results than that of millisecond activation of engram cells. Since during natural memory recall the memory trace is likely to be activated for only a few seconds to minutes, optogenetics appears to be a superior memory trace manipulation technique than DREADDs. One must keep in mind that engram cell manipulation experiments were a technological demonstration that c-Fos expressing cells that were activated during learning, continued to be involved in recall. Since studies that identify cells active during learning as well as recall do not show a complete overlap of cell populations (Guzowski et al., 1999; Reijmers et al., 2007; Vazdarjanova et al., 2002), the persistent tagging method employed by engram studies most likely over-represents the number and identity of hippocampal engram cells that would be involved in memory recall. Clearly, memory manipulation studies have focused on identifying and subsequently manipulating engram cells and have largely ignored the specific activity patterns central to Hebb's cell assembly hypothesis. This negligence is largely due to technical limitations associated with faithful replay of learning-associated activity patterns in live animals and future studies should address this issue.

In this chapter we also discussed the evidence that c-Fos expressing putative engram cells undergo synaptic plasticity after learning (Ryan et al., 2015). Further, we noticed that engram manipulation tools modulate the hippocampal synaptic strength. Yet some memory manipulation experiments employed repeated usage of cell manipulation technology which may have further induced synaptic plasticity, thereby complicating the interpretation of the data. Future studies will need to address this concern. Perhaps a study in which during the periods of artificial reactivations, synaptic plasticity is temporarily blocked, may resolve this issue.

While we specifically focused on the hippocampal memory trace that represents spatial and contextual information, a memory consists of different sensory experiences which are likely to be represented as distinct engrams distributed across several brain regions (see excellent reviews by (Josselyn et al., 2015; Tonegawa et al., 2015). We did not discuss how brain cells are allocated to a memory trace. This question has been previously addressed in the amygdalar neuronal networks and the data suggests that a neuron's current excitability state biases its chances of being recruited in a subsequent memory trace (Han et al., 2007; Josselyn et al., 2015). This allocation issue is also currently being probed in the hippocampus (Park et al., 2016). Further, engram allocation approach is also being used to test whether two memories that appear in close temporal proximity are represented by memory traces with overlapping populations of c-Fos positive cells (Cai et al., 2016; Rashid et al., 2016). The c-Fos tagging approach has also been used to verify the long-held assumption that place cells are memory cells (Andersen, 2007), by examining the similarities in experience-dependent changes in the spiking properties between c-Fos cell populations and place cell populations. Since some of this research is very recent, fine-tuning of the technology is necessary, replication of studies is required and differences in interpretations of these results will need to be sorted out before this new knowledge becomes part of text books and are established as facts.

References

Alexander GM, Rogan SC, Abbas AI, Armbruster BN, Pei Y, Allen JA, Nonneman RJ, Hartmann J, Moy SS, Nicolelis MA, et al. (2009) Remote control of neuronal activity in transgenic mice expressing evolved G protein-coupled receptors. *Neuron* 63: 27–39.

Andersen P (2007) *The hippocampus book* (Oxford University Press, USA).

Armbruster BN, Li X, Pausch MH, Herlitze S, and Roth BL (2007) Evolving the lock to fit the key to create a family of G protein-coupled receptors potently activated by an inert ligand. *Proc. Natl. Acad. Sci. U. S. A.* 104: 5163–5168.

Barth AL, Gerkin RC, and Dean KL (2004) Alteration of neuronal firing properties after in vivo experience in a FosGFP transgenic mouse. *J. Neurosci. Off. J. Soc. Neurosci.* 24: 6466–6475.

Bliss TV, and Lomo T (1973) Long-lasting potentiation of synaptic transmission in the dentate area of the anaesthetized rabbit following stimulation of the perforant path. *J. Physiol.* 232: 331–356.

Boyden ES, Zhang F, Bamberg E, Nagel G, and Deisseroth K (2005) Millisecond-timescale, genetically targeted optical control of neural activity. *Nat. Neurosci.* 8: 1263–1268.

Cai DJ, Aharoni D, Shuman T, Shobe J, Biane J, Song W, Wei B, Veshkini M, La-Vu M, Lou J, et al. (2016) A shared neural ensemble links distinct contextual memories encoded close in time. *Nature* 534: 115–118.

Cajal SRY (1894) The Croonian Lecture: La Fine Structure des Centres Nerveux. *Proc. R. Soc. Lond.* 55: 444–468.

Carr VA, Rissman J, and Wagner AD (2010) Imaging the human medial temporal lobe with high-resolution fMRI. *Neuron* 65: 298–308.

Chawla MK, Guzowski JF, Ramirez-Amaya V, Lipa P, Hoffman KL, Marriott LK, Worley PF, McNaughton BL, and Barnes CA (2005) Sparse, environmentally selective expression of Arc RNA in the upper blade of the rodent fascia dentata by brief spatial experience. *Hippocampus* 15: 579–586.

Chow BY, Han X, and Boyden ES (2012) Genetically encoded molecular tools for light-driven silencing of targeted neurons. *Prog. Brain Res.* 196: 49–61.

Denny CA, Kheirbek MA, Alba EL, Tanaka KF, Brachman RA, Laughman KB, Tomm NK, Turi GF, Losonczy A, and Hen R (2014) Hippocampal memory traces are differentially modulated by experience, time, and adult neurogenesis. *Neuron* 83: 189–201.

Dudek SM, and Bear MF (1992) Homosynaptic long-term depression in area CA1 of hippocampus and effects of N-methyl-D-aspartate receptor blockade. *Proc. Natl. Acad. Sci. U. S. A.* 89: 4363–4367.

Dunwiddie T, and Lynch G (1978) Long-term potentiation and depression of synaptic responses in the rat hippocampus: localization and frequency dependency. *J. Physiol*. 276: 353–367.

Epp JR, Silva Mera R, Köhler S, Josselyn SA, and Frankland PW (2016) Neurogenesis-mediated forgetting minimizes proactive interference. *Nat. Commun.* 7: 10838.

Fleischmann A, Hvalby O, Jensen V, Strekalova T, Zacher C, Layer LE, Kvello A, Reschke M, Spanagel R, Sprengel R, et al. (2003) Impaired long-term memory and NR2A-type NMDA receptor-dependent synaptic plasticity in mice lacking c-Fos in the CNS. *J. Neurosci. Off. J. Soc. Neurosci.* 23: 9116–9122.

Frankland PW, Köhler S, and Josselyn SA (2013) Hippocampal neurogenesis and forgetting. *Trends Neurosci.* 36: 497–503.

Garner AR, Rowland DC, Hwang SY, Baumgaertel K, Roth BL, Kentros C, and Mayford M (2012) Generation of a Synthetic Memory Trace. *Science* 335: 1513–1516.

Guzowski JF, McNaughton BL, Barnes CA, and Worley PF (1999) Environment-specific expression of the immediate-early gene Arc in hippocampal neuronal ensembles. *Nat. Neurosci.* 2: 1120–1124.

Guzowski JF, Setlow B, Wagner EK, and McGaugh JL (2001) Experience-dependent gene expression in the rat hippocampus after spatial learning: a comparison of the immediate-early genes Arc, c-fos, and zif268. *J. Neurosci. Off. J. Soc. Neurosci.* 21: 5089–5098.

Hall J, Thomas KL, and Everitt BJ (2001) Fear memory retrieval induces CREB phosphorylation and Fos expression within the amygdala. *Eur. J. Neurosci.* 13: 1453–1458.

Han JH, Kushner SA, Yiu AP, Cole CJ, Matynia A, Brown RA, Neve RL, Guzowski JF, Silva AJ, and Josselyn SA (2007) Neuronal competition and selection during memory formation. *Science* 316: 457–460.

Hebb DO (1949) *The Organization of Behavior: A Neuropsychological Theory* (L. Erlbaum Associates).

Hess US, Lynch G, and Gall CM (1995) Regional patterns of c-fos mRNA expression in rat hippocampus following exploration of a novel environment versus performance of a well-learned discrimination. *J. Neurosci. Off. J. Soc. Neurosci.* 15: 7796–7809.

Isaac JTR, Buchanan KA, Muller RU, and Mellor JR (2009) Hippocampal Place Cell Firing Patterns Can Induce Long-Term Synaptic Plasticity In Vitro. *J. Neurosci.* 29: 6840–6850.

Josselyn SA, Köhler S, and Frankland PW (2015) Finding the engram. *Nat. Rev. Neurosci.* 16: 521–534.

Lashley K (1950) In search of the engram. In *Physiological Mechanisms in Animal Behavior.* (Society's Symposium IV.), (Oxford, England: Academic Press), pp. 454–482.

Leutgeb JK, Leutgeb S, Treves A, Meyer R, Barnes CA, McNaughton BL, Moser MB, and Moser EI (2005) Progressive transformation of hippocampal neuronal representations in "morphed" environments. *Neuron* 48: 345–358.

Leutgeb JK, Leutgeb S, Moser MB, and Moser EI (2007) Pattern Separation in the Dentate Gyrus and CA3 of the Hippocampus. *Science* 315: 961–966.

Liu X, Ramirez S, Pang PT, Puryear CB, Govindarajan A, Deisseroth K, and Tonegawa S (2012) Optogenetic stimulation of a hippocampal engram activates fear memory recall. *Nature* 484: 381-385.

López AJ, Kramár E, Matheos DP, White AO, Kwapis J, Vogel-Ciernia A, Sakata K, Espinoza M, and Wood MA (2016) Promoter-Specific Effects of DREADD Modulation on Hippocampal Synaptic Plasticity and Memory Formation. *J. Neurosci. Off. J. Soc. Neurosci.* 36: 3588–3599.

Marr D (1971) Simple Memory: A Theory for Archicortex. *Philos. Trans. R. Soc. Lond. B Biol. Sci.* 262: 23–81.

Martin SJ, and Morris RGM (2002) New life in an old idea: The synaptic plasticity and memory hypothesis revisited. *Hippocampus* 12: 609–636.

Mayford M, Siegelbaum SA, and Kandel ER (2012) Synapses and Memory Storage. *Cold Spring Harb. Perspect. Biol.* 4: a005751.

McHugh TJ, and Tonegawa S (2007) Spatial exploration is required for the formation of contextual fear memory. *Behav. Neurosci.* 121: 335–339.

McHugh TJ, Blum KI, Tsien JZ, Tonegawa S, and Wilson MA (1996) Impaired hippocampal representation of space in CA1-specific NMDAR1 knockout mice. *Cell* 87: 1339–1349.

McNaughton BL, and Morris RGM (1987) Hippocampal synaptic enhancement and information storage within a distributed memory system. *Trends Neurosci.* 10: 408–415.

Morris RG (1989) Synaptic plasticity and learning: selective impairment of learning rats and blockade of long-term potentiation in vivo by the N-methyl-D-aspartate receptor antagonist AP5. *J. Neurosci. Off. J. Soc. Neurosci.* 9: 3040–3057.

Moser EI, Moser MB, and Andersen P (1994) Potentiation of dentate synapses initiated by exploratory learning in rats: dissociation from brain temperature, motor activity, and arousal. *Learn. Mem. Cold Spring Harb.* N 1: 55–73.

Nabavi S, Fox R, Proulx CD, Lin JY, Tsien RY, and Malinow R (2014) Engineering a memory with LTD and LTP. *Nature* 511: 348–352.

Nakashiba T, Cushman JD, Pelkey KA, Renaudineau S, Buhl DL, McHugh TJ, Rodriguez Barrera V, Chittajallu R, Iwamoto KS, McBain CJ, et al. (2012) Young dentate granule cells mediate pattern separation, whereas old granule cells facilitate pattern completion. *Cell* 149: 188–201.

Nakazawa K, McHugh TJ, Wilson MA, and Tonegawa S (2004) NMDA receptors, place cells and hippocampal spatial memory. *Nat. Rev. Neurosci.* 5: 361–372.

Nonaka A, Toyoda T, Miura Y, Hitora-Imamura N, Naka M, Eguchi M, Yamaguchi S, Ikegaya Y, Matsuki N, and Nomura H (2014) Synaptic plasticity associated with a memory engram in the basolateral amygdala. *J. Neurosci. Off. J. Soc. Neurosci.* 34: 9305–9309.

O'Keefe J, and Dostrovsky J (1971) The hippocampus as a spatial map. Preliminary evidence from unit activity in the freely-moving rat. *Brain Res.* 34: 171–175.

O'Keefe J, and Nadel L (1978) *The Hippocampus as a Cognitive Map* (Oxford University Press, USA).

O'Reilly RC, and Rudy JW (2001) Conjunctive representations in learning and memory: principles of cortical and hippocampal function. *Psychol. Rev.* 108: 311–345.

Park S, Kramer EE, Mercaldo V, Rashid AJ, Insel, N, Frankland, PW, and Josselyn, SA (2016) Neuronal Allocation to a Hippocampal Engram. *Neuropsychopharmacol. Off. Publ. Am. Coll. Neuropsychopharmacol.* 41: 2987–2993.

Ramirez S, Liu X, Lin PA, Suh J, Pignatelli M, Redondo RL, Ryan TJ, and Tonegawa S (2013) Creating a false memory in the hippocampus. *Science* 341: 387–391.

Rashid AJ, Yan C, Mercaldo V, Hsiang HL (Liz), Park S, Cole CJ, Cristofaro AD, Yu J, Ramakrishnan C, Lee SY, et al. (2016) Competition between engrams influences fear memory formation and recall. *Science* 353: 383–387.

Reijmers LG, Perkins BL, Matsuo N, and Mayford M (2007) Localization of a stable neural correlate of associative memory. *Science* 317: 1230–1233.

Ryan TJ, Roy DS, Pignatelli M, Arons A, and Tonegawa S (2015) Memory. Engram cells retain memory under retrograde amnesia. *Science* 348: 1007–1013.

Sanders J, Cowansage K, Baumgärtel K, and Mayford M (2012) Elimination of dendritic spines with long-term memory is specific to active circuits. *J. Neurosci. Off. J. Soc. Neurosci.* 32: 12570–12578.

Schacter DL, Eich JE, and Tulving E (1978) Richard Semon's theory of memory. *J. Verbal Learn. Verbal Behav.* 17: 721–743.

Scoville WB, and Milner B (1957) Loss of recent memory after bilateral hippocampal lesions. *J. Neurol. Neurosurg. Psychiatry* 20: 11–21.

Semon RW (1921) *The mneme* (New York, NY: Allen & Unwin Limited).

Takata N, Yoshida K, Komaki Y, Xu M, Sakai Y, Hikishima K, Mimura M, Okano H, and Tanaka KF (2015) Optogenetic activation of CA1 pyramidal neurons at the dorsal and ventral hippocampus evokes distinct brain-wide responses revealed by mouse fMRI. *PloS One* 10: e0121417.

Takeuchi T, Duszkiewicz AJ, and Morris RGM (2014) The synaptic plasticity and memory hypothesis: encoding, storage and persistence. *Philos. Trans. R. Soc. B Biol. Sci.* 369(1633): 20130288.

Tanaka KZ, Pevzner A, Hamidi AB, Nakazawa Y, Graham J, and Wiltgen BJ (2014) Cortical representations are reinstated by the hippocampus during memory retrieval. *Neuron* 84: 347–354.

Thompson LT, and Best PJ (1990) Long-term stability of the place-field activity of single units recorded from the dorsal hippocampus of freely behaving rats. *Brain Res*. 509: 299–308.

Tomar A, Polygalov D, Chattarji S, and McHugh TJ (2015) The dynamic impact of repeated stress on the hippocampal spatial map. *Hippocampus 25*(1): 38-50.

Tonegawa S, Liu X, Ramirez S, and Redondo R (2015) Memory Engram Cells Have Come of Age. *Neuron* 87: 918–931.

Vann SD, Brown MW, Erichsen JT, and Aggleton JP (2000) Fos imaging reveals differential patterns of hippocampal and parahippocampal subfield activation in rats in response to different spatial memory tests. *J. Neurosci. Off. J. Soc. Neurosci.* 20: 2711–2718.

Vazdarjanova A, McNaughton BL, Barnes CA, Worley PF, and Guzowski JF (2002) Experience-dependent coincident expression of the effector immediate-early genes arc and Homer 1a in hippocampal and neocortical neuronal networks. *J. Neurosci. Off. J. Soc. Neurosci.* 22: 10067–10071.

Vinogradova OS (2001) Hippocampus as comparator: role of the two input and two output systems of the hippocampus in selection and registration of information. *Hippocampus* 11: 578–598.

Wang KH, Majewska A, Schummers J, Farley B, Hu C, Sur M, and Tonegawa S (2006) In vivo two-photon imaging reveals a role of arc in enhancing orientation specificity in visual cortex. *Cell* 126: 389–402.

Whitlock JR, Heynen AJ, Shuler MG, and Bear MF (2006) Learning induces long-term potentiation in the hippocampus. *Science* 313: 1093–1097.

Wills TJ, Lever C, Cacucci F, Burgess N, and O'Keefe J (2005) Attractor dynamics in the hippocampal representation of the local environment. *Science* 308: 873–876.

Wilson MA, and McNaughton BL (1993) Dynamics of the hippocampal ensemble code for space. *Science* 261: 1055–1058.

Wilson MA, and McNaughton BL (1994) Reactivation of hippocampal ensemble memories during sleep. *Science* 265: 676–679.

Wilson Y, Nag N, Davern P, Oldfield BJ, McKinley MJ, Greferath U, and Murphy M (2002) Visualization of functionally activated circuitry in the brain. *Proc. Natl. Acad. Sci. U. S. A.* 99: 3252–3257.

Zhang J, McQuade JMS, Vorhees CV, and Xu M (2002) Hippocampal expression of c-fos is not essential for spatial learning. *Synapse* 46: 91–99.

Zhu H, Pleil KE, Urban DJ, Moy SS, Kash TL, and Roth BL (2014) Chemogenetic Inactivation of Ventral Hippocampal Glutamatergic Neurons Disrupts Consolidation of Contextual Fear Memory. *Neuropsychopharmacology* 39: 1880–1892.

In: Synaptic Plasticity
Editor: Sraboni Chaudhury

ISBN: 978-1-53613-606-7

Chapter 10

TOOLS AND TECHNIQUES TO STUDY SYNAPTIC PLASTICITY

***Vivek Kumar*[1,*] *and Shilpi Singh*[2]**
[1]Molecular and Behavioral Neuroscience Institute,
University of Michigan, Ann Arbor, MI, US
[2]Amity Institute of Biotechnology,
Amity University, Mumbai, Maharashtra, India

ABSTRACT

Synaptic plasticity serves as the most fundamental mechanism of learning and memory. A wide array of tools and techniques has been employed to study the changes in the synaptic strength which mainly occurs as long-term potentiation (LTP) or long-term depression (LTD). Traditionally, researchers have used different biochemical assays and electrophysiological recordings to investigate the synaptic changes. Biochemical assays and genetic manipulations have allowed to study the energy and calcium metabolism, neurotransmitter synthesis, transport and secretion, and downstream signaling mechanisms. On the other hand, electrophysiological techniques allow real time monitoring of

* Corresponding Author: vkmbni@umich.edu.

bioelectrical activity from various models including neuromuscular junctions, single neurons, tissue slices and *in vivo* animal or human brain through intracellular or extracellular recordings. These classical methods have contributed significantly to our knowledge of neuronal activity, role of biochemical pathways in the induction and expression of plasticity, mechanisms of LTP/LTD and memory formation. The major limitation- spatial and temporal resolutions of synaptic events that follow any synaptic stimulation, of these classical approaches are circumvented by the recent discovery and advances in the optogenetics and optical imaging. Optogenetics involves genetic methods to deliver opsin genes to express light-activated channels within targeted neuron populations allowing these neurons to be selectively activated or inhibited by different wavelengths of light. It is applicable for studying neuronal activity, *in vitro* and *in vivo* synaptic plasticity and holds a strong future to elucidate the learning and memory formation mechanisms with high spatiotemporal precision. To observe these synaptic or associated cellular changes whether at molecular or morphological level, imaging techniques play vital role. Different forms of fluorescence microscopy including wide field microscopy, confocal microscopy and two-photon microscopy combined with the use of fluorescent biosensors (e.g., FM dyes and Ca2+ sensitive dyes) has significantly increased our ability to study activity-dependent changes in cellular signaling across space and time and provide new insights into the mechanistic basis of synaptic plasticity.

Keywords: Electrophysiology, Hippocampus, Long-term potentiation, Patch-clamp technique, fEPSP, Optogenetics, ChR2-YFP, Fluorescent biomarker, Ca 2+ sensitive dyes, FM (FM1-43) dye, FRET, FLIM, Confocal microscopy, Two photon microscopy, PALM, STORM, Super-resolution imaging, SSIM, STED

1. INTRODUCTION

Synaptic plasticity is the principal focus of the modern neuroscience research and it results from a wide range of biochemical, physiological and structural changes within a neuron following any chemical or electrical stimulation. Biochemical (e.g., gel electrophoresis, western blotting, immunohistochemistry) (Kumar et al., 2014; Sanyal et al., 2013;

Choudhury et al., 2010) and genetic (e.g., Cre-lox based conditional gene targeting or viral vector-directed gene expression to generate transgenic and knock-out animals) (Cui et al., 2005) approaches have served as the first line of tools to help identify and assign the functions of genes and proteins essential for initiation and maintenance of synaptic plasticity events within individual neurons or an identified brain circuits or region. Along with biochemical assays, electrophysiological recordings have been used to investigate long-term changes in synaptic efficacy following the electrical, pharmacological, or genetic manipulation of synaptic function. However, these methodologies limit either the spatial or temporal resolution with which biological processes can be observed or manipulated. For example, the electrophysiological characterization of long-term potentiation (LTP), as reported by an increases in the amplitude of evoked postsynaptic potentials, often represents a collective change in transmission efficacy across a population of synapses rather than a direct characterization of plasticity at single sites. Moreover, the use of molecular and pharmacological techniques, although useful in elucidating the role of biochemical pathways in the induction and expression of plasticity, reveal little of the spatiotemporal dynamics of cellular signaling that follow synaptic stimulation. Such dynamics, however, are important in understanding how synaptic processing is altered across space and time following the induction of LTP. In the past decade, advances in optical imaging, combined with the genetic approaches like optogenetics and development of fluorescent biosensors, have circumvented the problems associated with traditional experimental techniques by offering unparalleled spatiotemporal resolution of molecular events at individual synapses and circuits of interest. Current chapter reviews the principles and applications of selected techniques (electrophysiology, optogenetics, fluorescent biomarkers, and optical imaging) widely used to study synaptic plasticity.

2. ELECTROPHYSIOLOGY

Bioelectrical activity in nervous system, since its first discovery by Galvani in 1971 (Piccolino, 1998), is considered as the functional basis of the animal behavior. By using various electrophysiological techniques to monitor in real time, researchers have been trying to unravel the mechanisms of bioelectrical activity - generation, conduction and its coordination at single cell level to form a functional neural circuitry. Eventually, these methods have also contributed significantly to understand the mechanisms of several pathological conditions associated with abnormal electrical activity (e.g., epilepsy) and synaptic plasticity which is a critical phenomenon for learning and memory. Dysregulation in synaptic plasticity leads to several cognition deficient conditions like Alzheimer's disease (Arendt, 2001; Arendt et al., 1991).

In order to study the synaptic development and function using electrophysiology, the simplest and precisely tractable model is neuromuscular junction (NMJ), especially the Drosophila larval NMJ. Features like robust plasticity, advanced Drosophila genetics which allows precise manipulation of circuits, and glutamatergic nature (unlike the vertebrate cholinergic NMJ) of Drosophila NMJ, makes it a highly favorable tool to study synaptic plasticity (Collins and DiAntonio, 2007). One can use a variety of genetic tools to – 1) target the directed expression of modified ion channels, 2) alter the synaptic transmission rate (quantal content or quantal size), 3) control the expression levels of secondary messengers (e.g., cAMP, CAMKII), growth factors (e.g., TGFb), or transcription factors (e.g., Fos/Jun, CREB) to manipulate the membrane excitability to study the synaptic plasticity (Olsen and Keshishian, 2012).

At tissue level, one brain region whose anatomical uniformity and critical role in learning and memory, made it highly suitable for electrophysiological studies, is the hippocampus. A brief and strong electrical stimulation can induce hippocampal synaptic function gain which may last for hours to days (Bliss and Lomo, 1973). Such an induction of long-lasting change in synaptic strength is termed long-term potentiation (LTP). The magnitude of LTP, or synaptic plasticity which reflects the collective depolarization and excitation of a significant number of neurons, represents cognitive function or memory consolidation (Bliss and Collingridge, 1993). Using neurophysiological study, studies have proved that by augmenting the synaptic function through induction of LTP in naïve animals, one can enhance the memory and cognition (Lynch and Gall, 2013; Lynch et al., 2011), and at the same time, the investigations of deficits in the magnitude of LTP correlates with the disease conditions (Nisticò et al., 2012).

Electrophysiological studies can be performed through either intracellular or extracellular recordings.

a) Intracellular recordings allow one to measure direct transmembrane potentials (difference in electric potential between the cell membrane and the extracellular environment), or current in individual neurons. Several modes of intracellular recordings have been established based on the requirement of whether one wants to measure voltage and/or current across the membrane of a cell.
One mode of intracellular recording, which has greatly facilitated the investigation of synaptic currents, is the "patch-clamp technique". This intracellular recording method can be performed in three main configurations: 1) cell attached: a glass microelectrode is placed next to the cell membrane and using gentle suction, a small piece of cell membrane is drawn into the micropipette tip. Using this mode one can study single ion channels present on the patch of the membrane. It has allowed the discovery of new classes of ion channels and their physiological roles in not only the excitable neurons but also in not electrically excitable neurons 2) whole cell: with little extra suction, the cell membrane patch gets displaced, allowing the electrode solution to come in direct contact with the cytosol which in turn allows very stable intracellular recording. This permits the investigation of ion channel conductance and kinetic behavior. 3) perforated patch: highly suitable for pharmacological studies as it allows the benefits of both the above mentioned modes. Once a patch is formed, pore-

forming agents like antibiotics are used to generate micropores which allows ionic continuity between cytosol and electrode solution but prevents intracellular proteins from being exchanged across the solutions (Conforti, 2012; Kornreich, 2007). This method can be used to measure synaptic currents or potential from single neurons in culture (Tong et al., 1996) or tissue slice preparations (Manabe, 2012; Manabe et al., 2000) as well as *in vivo* in awake rodents (Lee and Lee, 2017). Intracellular recordings can be done in either voltage clamp or current clamp mode (Johnston et al., 1988). For studying synaptic functions involving voltage-gated channels, one can use the voltage clamp technique which allows to measure ionic current while keeping the voltage constant across the membrane (Cole, 1949; Hamill et al., 1981). Using voltage clamp technique, hippocampal mossy fiber synapses show enhanced LTP following noradrenergic stimulation mediated via cAMP pathway. (Johnston et al., 1988).

Current clamp technique is another mode, which provides researcher an opportunity to record membrane potential by injecting a fixed amount of current into a neuron and allows to answer the most basic question of how excitable neurons react to the incoming electric current or neurotransmitters which in turn act by opening ion channels and initiating current flow. Using current clamp study, it has been shown that one can induce global plasticity in apical but not basal dendrites of CA1 pyramidal neurons through pathway-specific modulation and such a global LTP was generated exclusively by postsynaptic burst activity only in CA1 (Berberich et al., 2017). These electrophysiological modes have also confirmed that unlike CA1, the two other regions of hippocampus-CA2 and CA3 behave differently in terms of synaptic plasticity. Studies on hippocampal plasticity have shown that CA2 and CA3 neurons possess distinctive electrophysiological features that include a unique synaptic stability where neurons fail to show any significant LTP or long term depression (LTD) induction (Berberich et al., 2017; Zhao et al., 2007). This could help us to understand the mechanisms by which CA2/CA3 regions show an appreciable resistance to damage from adverse synaptic events like disease or hypoxic conditions.

b) Extracellular recordings allow measurements of the variation in the electrical potential or electric discharge in close vicinity of either a single neuron (single-unit recording) or an identified group of neurons (multi-unit recording). Since this technique relies on recordings from neurons without stressing them, a single neuron or region can be studied for longer durations either in slice preparation or live animals. Hence, it allows one to comprehensively characterize physiology and connectivity of individual neurons or a groups of neurons and in turn can be correlated with the behavior of the animal.

   Apart from, single- or multi-unit recordings which measure spike discharges from neurons or groups of neurons respectively, once can also measure local synchronized currents resulting from the simultaneous activation of group of neurons in a given region of brain. The latter technique is referred as extracellular filed potential measurement (Abrahamsson et al., 2016). By measuring field excitatory postsynaptic potential (fEPSP) in rat hippocampal slices, reports show that formalin-induced persistent pain significantly increased the number of detectable fEPSP in both DG and CA1 area, indicating that peripheral persistent nociception can affect the temporal and spatial plasticity of synaptic connection and function in the hippocampal formation (Zhao et al., 2009). In an attempt to understand the link between synaptic activity and Alzheimer's disease, a study found that hippocampal seizure can significantly increase the inter-synaptic fluid amyloid beta levels by using mice hippocampal slice preparations and extracellular filed potential recordings. Such an increased synaptic activity can be reduced by providing activity blockers like Tetrodotoxin (Cirrito et al., 2005).

   Measurement of extracellular field potential following the induction of white-matter-evoked LTP in visual cortical slices at different ages from light-reared and dark-reared rats showed that LTP is an experience-dependent synaptic modification in the developing mammalian brain and co-regulated by age and experience (Kirkwood et al., 1995).

3. Optogenetics

Optogenetics is a recently developed technique which enables non-invasive optical control of specific neurons with milliseconds precision (Boyden et al., 2005; Zhang et al., 2006). It allows targeted and selective control of neuronal activity among a complex population and thus overcomes the major limitation of electrophysiological approach and can be easily applied to freely behaving live animals for *in vivo* studies (Johansen et al., 2010; Murakoshi et al., 2017). Optogenetics relies on genetic methods to deliver microbial opsin genes to express light-activated channels (similar to those found in retinal photoreceptors/rhodopsin) within targeted neuron populations allowing these neurons to be selectively activated or inhibited by different wavelengths of light. It is applicable for studying neuronal activity, *in vitro* and *in vivo* synaptic plasticity and holds a strong future to elucidate the learning and memory formation mechanisms with high spatiotemporal precision (Ben-Simon et al., 2015; Gradinaru et al., 2010; Zhang et al., 2010).

3.1. *In Vitro* Study

Brain slices containing areas of interest has proved to be a vital and easy model to perform *in vitro* studies for synaptic plasticity. Using optogenetic-field potential recordings in Channelrhodopsin-2-Yellow fluorescent protein (ChR2-YFP) expressing transgenic mice brain slice preparations, results show that extracellular field potentials could be reliably induced and stably maintained for longer duration in layer II/III and layer V of neocortical slices (Xiong and Jin, 2012). Using single-cell electrophysiology and optogenetics in thalamocortical (TC) slices containing the auditory thalamus and auditory cortex (ACx), LTP is expressed at TC synapses in both young and mature mice but gets masked by certain gating mechanism after early critical period. This postsynaptically expressed TC LTP, dependent on group I metabotropic glutamate receptors, can be unmasked in mature ACx by cortical disinhibition combined with activation of M1 muscarinic inputs from the nucleus basalis. These long lasting TC LTP in the ACx can potentially contribute to experience-dependent cortical map plasticity in the ACx in both young and adults (Chun et al., 2013). Similarly, single pulses of septal cholinergic input, activated either by electrical stimulation or via optogenetically, induces LTP or LTD, in hippocampal Schaeffer collateral (SC) to CA1 synaptic plasticity which depends upon the timing (100 or 10 msec, respectively) of cholinergic input prior to the SC input. Such observations reveal not only the temporal advantages of optogenetics but also the remarkable temporal precision of cholinergic transmission and involvement of the hippocampus and cholinergic transmission in higher cognitive functions (Gu and Yakel, 2011).

Optogenetics also provides a clear advantage to study spatial aspect of synaptic plasticity like asymmetry at synaptic locations. A recent study showed that CA1 synapses receiving neuronal input from the right CA3 pyramidal cells are larger and have more perforated PSD and a GluR1 expression level twice as high as those receiving input from the left CA3. It also showed that there is a unique correlation between synapse size and the expression of different glutamate receptor subunits e.g., synaptic density of GluR1 increases as the size of a synapse increases (Shinohara et al., 2008). In addition, photostimulation of astrocytes expressing ChR2 produces LTD in Purkinje cell of cortical slices by activating metabotropic glutamate receptors (Sasaki et al., 2012), which further expands the range of applications for optogenetics in synaptic plasticity.

3.2. *In Vivo* Study

Use of optogenetics to explore plasticity under *in vivo* conditions has enormous potential to provide a wealth of information. For example, stimulation of dopaminergic neurons in the ventral tegmental area (VTA) that express ChR2 facilitates the development of positive reinforcement during reward-seeking and of behavioral flexibility in freely moving mice (Adamantidis et al., 2011). Another study used optogenetics to understand the cocaine evoked plasticity and potential therapeutics development for addiction. This study showed that cocaine via ERK signaling, can potentiate excitatory transmission in D1-receptor-expressing medium-sized spiny neurons (D1R-MSNs) in mice, which can be reversed (abolishment of cocaine-induced locomotor sensitization) by the optogenetic stimulation of cortical nucleus accumbens inputs. These findings hint towards a selective potentiation in D1R-MSNs as a mechanism underlying a core component of addiction and provide a novel basis for deep brain stimulation or transcranial magnetic stimulation (Pascoli et al., 2011). Another interesting study showed that photostimulation with moderate light sequence to depolarize the ChR2 expressing Bergmann glial cells in mouse, can alter the motor behavioral (measured as perturbation of the smooth eye pursuit of the visual stimuli) plasticity modulated by the cerebellum (Sasaki et al., 2012).

Furthermore, optogenetics has already provided critical results on plasticity from the brain regions related to memory-storage. A recent study using optogenetic manipulation of hypocretin/orexin (Hcrt) neurons (key role in arousal processes) in lateral hypothalamus revealed the role of sleep continuity in the memory consolidation. Authors observed an impairment of memory consolidation (novel object recognition task) when ChR2 expressing Hcrt cells were optically stimulated at 60-s intervals and in turn led to sleep fragmentation (Rolls et al., 2011). Another study used optogenetic manipulations to alter activity after a learning event to investigate the processes underlying memory consolidation. This study reported that stimulation of ChR2 (E123A) expressing neurons in the basolateral amygdala with 40-Hz light pulses enhanced memory retention (inhibitory avoidance task) whereas inhibition of archaerhodopsin-3-expressing neurons in the basolateral amygdala for 15 min, significantly impaired retention (Huff et al., 2013). The following study using optogenetics contributed to our understating about spatial and temporal factors controlling memory retention and involvement of hippocampus and other structures. By stimulating the fast optogenetic inhibitor eNpHR3.1 expressing neurons in dorsal hippocampus to inhibit the CA1 excitatory neurons, authors showed that it is possible to block the acquisition and retrieval of contextual fear memory and reversibly impair the remote fear memory recall which can only be done by a precise but not prolonged inhibition (Goshen et al., 2011). Another excellent study utilized optogenetics to investigate whether a light-activated contextual memory in the DG or CA1 can serve as a functional conditioned stimulus (CS) in fear conditioning. Results showed that the manipulation of ChR2-expressing neurons in the dentate gyrus but not in the ChR2-expressing neurons of CA1, can subsequently serve as a functional CS in a fear-conditioning paradigm when artificially reactivated during the delivery of an unconditioned stimulus (US). The consequence is the formation of a false associative fear memory to the CS that was not naturally available at the time of the US delivery (Ramirez et al., 2013). Taken together, these studies suggest that memory formation is spatiotemporally precise and advanced and precise techniques like optogenetics can provide highly valuable information for the plasticity mechanisms and memory formation under *in vivo* conditions.

Major limitations of ChR2 optogenetics include -1) activation of voltage-gated calcium channels by photostimulation (Zhang and Oertner, 2007), 2) temperature sensitive kinetics of ChR2, and 3) effects of depolarization (Chater et al., 2010). Minor concerns include - behavioral interference from intrinsic phototactic reactions induced by ionic exchanges through opsin channels and phototoxicity due to the long-term illumination, although these have been successfully shown to overcome by using the ChR2 mutant C128X (Schultheis et al., 2011). Similar improvements and availability of more suitable opsins will allow multiplexing of light-stimulated channels and thus more precise control of specific neuronal populations.

4. Fluorescent Biosensors and Optical Imaging

Apart from the biochemical and electrical mechanisms involving neurotransmitters and receptors, synaptic plasticity is also dictated by the morphological changes in the synapses including- structure, size or number of the synapses constituting the structural plasticity. Technical advances in microscopy accompanied by development in the fluorescent protein biochemistry and its targeted expression, and fluorescent dyes, has made it easier and more precise to image individual synapses over extended periods of time in the brain at high resolution in living animals. Such an approach has revealed fundamental insights into the plasticity of synapses in the peripheral nervous system and their sculpting over development. Following sections discusses the availability, use and advancement of fluorescent biomarker and optical techniques which can be employed to study synaptic plasticity.

4.1. Fluorescent Marker Biosensors

4.1.1. LTP at Presynaptic Side

Presynaptic components of LTP can be directly investigated through optical assays of vesicular fusion by using Ca 2+ sensitive dyes and FM. Fluorescent Ca2+ indicator dyes allow one to assess the neurotransmitter release at single-synapse resolution and helped to establish the fact that induction of LTP is triggered by an elevation of postsynaptic Ca2+ initiated by activation of NMDA receptors (Emptage et al., 1999; Lynch et al., 1983; Malenka et al., 1988). Ca2+ transients can be generated at stimulated spines by single-pulse excitation of afferent fibers. Since these transients result from Ca2+ influx via NMDARs and voltage-gated channels but depend on AMPA-receptor activation, represent the presynaptic release (Emptage et al., 1999; Emptage et al., 2003; Reid et al., 2001). By monitoring the probability and amplitude of generated excitatory postsynaptic Ca2+ transients simultaneously, one can measure the efficiency of both pre- and post-synaptic transmission, respectively (Reid et al., 2004; Reid et al., 2001). Studies using only Ca2+ imaging have revealed that tetanic stimulation evokes transients at Schaffer-collateral synapses (Emptage et al., 1999; Emptage et al., 2003; Ward et al., 2006) but owing to a non-linear relation with AMPA receptors, this technique is not sufficient to assess postsynaptic activity following LTP (Emptage et al., 1999). In order to simultaneously study both pre and post synaptic components of LTP at Schaffer-collateral synapses, one can combine Ca2+ imaging with electrophysiological recordings (Enoki et al., 2009).

FM (FM1-43) dye is a fluorescent amphipathic styryl dye, which can easily intercalate with the plasma membrane (Betz and Bewick, 1992) and up taken by the recycling synaptic vesicles. Activity dependent unloading of FM 1-43 provides a direct measure of presynaptic function and has been successfully used to quantitate individual synaptic function even in high throughput experiment (Murthy et al., 1997; Ryan et al., 1993; Zakharenko et al., 2001). Chemically or high-frequency induced elevated extracellular K+ triggers loading of synaptic vesicles with Dye, leading to influx of most of the extracellular/extrasynaptic/cell membrane bound FM dye, leaving only intracellular vesicles labelled. Next round of synaptic activity or low frequency stimulation causes vesicular fusion and dye unloading in synaptic cleft which is detected as a decrease in fluorescence at individual synaptic boutons; the magnitude of which represents the presynaptic efficacy (Betz and Bewick, 1992). FM-imaging in combination with electrophysiological recordings, has revealed the induction mechanisms required for the pre- and post-synaptic expression of LTP. Although, the enhanced FM-destaining/unloading rates indicate an induced presynaptic change in the Schaffer-collaterals following high-frequency stimulation or chemical LTP induction but expression of full plasticity requires the activation of L-type voltage-gated calcium channels (L-VGCC) and NMDA-receptors (NMDAR) and presynaptic release of BDNF (Blundon and Zakharenko, 2008; Zakharenko et al., 2001). Once can also confirm such pharmacological and electrophysiological dissociation of the pre- and post-synaptic induction of LTP by using pHlourin-tagged synaptophysin (synaptopHlourin) and by monitoring vesicle fusion (Bayazitov et al., 2007).

4.1.2. LTP at Postsynaptic Side

The postsynaptic expression of LTP mainly involves an increase in the number of AMPA receptors at the synapse which is reflective of an activity-driven capture of lateral receptors at the synapse. Synaptic recruitment of AMPA receptors which usually takes place through either intracellular pool, via exocytosis (Cottrell et al., 2000; Man et al., 2000) or by lateral diffusion of extrasynaptic receptors, can be studied using single particle tracking (SPT), via fluorescent quantum dot tagged antibodies (Choquet, 2010; Dahan et al., 2003; Triller and Choquet, 2005; Weiss, 1999), or superecliptic pHlourins (SEP), pH-sensitive variant of green fluorescent protein (GFP). Both of these methods rely on the quenching of fluorochromes inside the acidic environment of endocytic vesicles but become fluorescent when present on the cell surface. Following the induction of LTP, use of SPT and SEP has shown a short-lived increase of synaptic AMPA receptor mobility and a simultaneous decrease in the number of receptors in the surrounding region. This leads to an increased proportion of immobile synaptic receptors (Tardin et al., 2003), but immediately this mobile pool of surface receptors is replenished by activity-dependent exocytosis (Makino and Malinow, 2009). In contrast to LTP, LTD leads to an increased receptor mobility in and around the synapse indicating the diffusion of immobilized receptors out of the synapse (Tardin et al., 2003). However, SPT is restricted to neuronal culture and responsive only to global modifications of synaptic activity and may not necessarily indicate an *in vivo* synaptic plasticity whereas, SEP requires overexpression of AMPA receptors which might affect the overall receptor dynamics and plasticity.

4.1.3. Synaptic Signaling

Understanding the spatiotemporal dynamics of synaptic signaling has gained much pace with the development of new optical probes and techniques like fluorescence resonance energy transfer (FRET)/ fluorescence lifetime imaging microscopy (FLIM) (Lakowicz, 2006; Miyawaki, 2003; Miyawaki et al., 1997; Okamoto et al., 2004; Wang and Taylor, 1981). FRET results when two genetically encoded fluorophores interact in close spatial proximity (<100Å), allowing one to study vital signaling events in spatiotemporal manner, monitoring the distribution and structure of proteins or protein-protein interactions. Spectral profile of the observed signal depends upon the proximity of the two fluorophores which are sensitive to the changes in protein confirmation or protein-protein interaction. The difficulty to measure the FRET signal which highly depends on the concentration of the reporter, can be overcome by using FLIM. This technique relies on the length of time required for fluorescence signals to decay and is, therefore, independent of reporter concentration (Murakoshi et al., 2008). FRET/FLIM-based reporters have been used to study the activation of CAMKII (Lee et al., 2009; Takao et al., 2005), ERK (Harvey et al., 2008), Ras (Yasuda et al., 2006), RhoA and Cdc42 (Murakoshi et al., 2011) and calpain (Vanderklish et al., 2000) and helped to understand the compartmentalized synaptic signaling. For example, single spine potentiation induced LTP resulted in transient activation of CAMKII, RhoA and Cdc42 which were restricted to the stimulated spine (Lee et al., 2009). In principle, one can design a vast number of genetically-encoded optical detector using this method. Already several *in vivo* studies have applied this technique to monitor activity-dependent Ca2+ dynamics, a key mechanism to understand plasticity (Tian et al., 2009).

4.2. Optical Imaging Systems

Fluorescence microscopy has advanced as a major tool to study neuronal physiology. It's not just limited to the co-localization and interaction studies, but can also be used to determine ion concentrations, as well as intra- and intercellular processes like endocytosis and exocytosis, and distribution of a single molecule species. With the recent development of high performance optical components including electro-optics, fluorescent biosensors, and high precision computerized control, one can now even perform distribution and dynamics of single molecules with biological tissues using super-resolution microscopy. Some of the most common and widely used variants of modern fluorescence microscopy and how these can be used to study synaptic plasticity, are discussed below.

4.2.1. Laser Scanning Confocal Microscopy

Laser scanning confocal microscopy (LSCM) is perhaps, the most widely used fluorescent microscopic system which not only allows multiplex detection but also the use of thick tissue specimen. The main principle of the confocal microscopy is to remove any out-of-focus light from reaching the detector with the help of a pinhole aperture. This ensures that light reaching the detector belongs to the same focal point in the specimen where the excitation light was focused. Using oscillating mirrors, confocal excitation and detection point across the specimen are scanned and compiled to generate an image. Since, its first introduction by Minsky in 1957 and application to neuroscience in 1967 (Egger and Petrăn, 1967), LSCM has gained a prominent role in its ability to provide unprecedented spatiotemporal details of synaptic structure and function. Study on the frog neuromuscular junction (NMJ) using omega-conotoxin GVIA, an irreversible Ca2+ channel ligand, and LSCM, revealed that the distribution of calcium channels is not random but localized at presynaptic membrane adjacent to the postsynaptic membrane (Robitaille et al., 1990). Visualization of large synaptic boutons of the lizard neuromuscular junction using LSCM revealed that intracellular calcium concentration (major determinant of neurotransmitter release) inside the presynaptic bouton are not constant even at rest but fluctuates from scan to scan (Rahamimoff and Melamed, 1993). Fluorescent lipophilic dye-Dil was used in combination with LSCM, to develop an optical method to visualize and quantify the synaptic structures in living brain tissue (Hosokawa et al.,

1994). Although, confocal microscope improves image resolution by eliminating the out-of-focus light but has limited depth penetration in thicker tissues, which is mainly caused by the scattering and absorption of near-by or out of focus tissue components.

4.2.2. Two Photon Laser Scanning Microscopy

Two photon microscopy (TPM) allows one to perform deep tissue imaging in live animal with minimal signal loss, photobleaching, or phototoxic damage (Helmchen and Denk, 2005). Instead of using a single photon like confocal, it depends on nearly simultaneous absorption of two photons by a single fluorophore. Most of the *in vivo* TPM imaging studies have investigated the dendritic spine elimination and formation. Repeated imaging over several weeks in the adult mouse (4 months of age) primary visual cortex showed remarkably stable spines in apical dendrites from layer V pyramidal neurons whereas in mice between 1 to 3 months of age, the significantly higher rate of spine elimination leads to a net loss of spines during young adolescence (Grutzendler et al., 2002). Zuo et al., has shown that sensory deprivation through whisker trimming, during second postnatal month preferentially reduces the spine loss, but not he spine formation (Zuo et al., 2005). However, such a deprivation has very minimal effects of spine dynamics as animals reach adult stage. Degradation of synaptic structure and function is one of the early features in many neurodegenerative diseases including Alzheimer's (AD) and cerebral ischemic diseases. Study performed by the Tsai et al., using *in vivo* TPM on a transgenic model of AD, has shown that deposition of amyloid triggers continuous remodeling of spines and dendrites, eventually leading to large-scale disruption of neuronal connections. (Tsai et al., 2005). Another example of studying induced hippocampal plasticity by using two-photon time-lapse microscopy and imaging the dendritic morphology of GFP-positive CA1 pyramidal neurons in organotypic hippocampal slices, reported that low-frequency stimulation (stimulates LTD) induced NMDA receptor-dependent spine loss, whereas theta burst stimulation (stimulates LTP) led to the formation of new spines (Nägerl et al., 2004). A modified version of TPM - two-photon super-resolution patterned excitation reconstruction microscopy was used to show an increase in dendritic spine head and neck width sizes on CA1 pyramidal neurons following theta-burst stimulation of Schaffer collateral axons (Urban et al., 2017).

4.2.3. Wide Field Microscopy

Although, TPM and CLSM are most widely used fluorescent optical systems and have been used extensively to study spine morphology (typically 0.2 to 2 um in length), but because of the limitation with lateral resolution (TPM = >250 nm, CLSM >200 nm), they prove to be inadequate to fully resolve finer structures like spine necks (Ding et al., 2009; Nägerl et al., 2008).

Recent and exciting development in wide field optical microscopy has well addressed the issue of sub-diffraction level resolution and led to the establishment of a brand new sub-field "super resolution microscopy." A number of variations has been developed which has already proved to be efficient tools to analyze the molecular architecture of sub-cellular structures in tissue samples with nanometer precision and promise to push forward the investigation of structural functional analysis.

4.2.3.1. Photoactivated Localization Microscopy (PALM)/Stochastic Optical Reconstruction Microscopy (STORM)

In principle, PALM and STORM separate the fluorescence of individual fluorophore by temporally separating their emission times, leading to the optical imaging of even intracellular proteins at nanometer spatial resolution. Although these two methods provide the opportunity to acquire the best spatial resolution but have very limited temporal resolution. PALM utilizes photoactivatable genetically encoded proteins fluorescent protein molecules in low density, which are activated, localized (to approximately 2 to 25 nanometers), and then bleached and finally, the collective position information from all fluorophores is assembled to generate a super-resolution image (Betzig et al., 2006). Several studies have shown the application of PALM to super-resolution live-cell imaging of adhesion complexes—transmembrane cytoskeleton-substrate attachment points central to the process of cell migration (Shroff et al., 2008).

On the other hand, STORM utilizes photo-switchable fluorescent organic dyes which are switched on and off using light of different colors and a fluorescence image is constructed from high-accuracy localization of individual fluorescent molecules (Rust et al., 2006). STORM based multicolor and 3D super-resolution imaging can be used to quantitatively characterize the spatial relationship of different proteins within the synapse and to determine the relative orientation of large molecular species (Dani et al., 2010). Adish Dani and his group also reported that stimulation of the vomeronasal organ (VNO) neurons in the ChR2-YFP expressing mice caused activity dependent synaptic changes in the accessory olfactory bulbs. They observed a significant synapse maturation effect - shift in the NMDA to AMPA receptor composition, due to the VNO stimulation and such an activity dependent plasticity is known to provide a postsynaptic paradigm for long-term potentiation during postnatal development (Kerchner and Nicoll, 2008).

4.2.3.2. Saturated Structured-Illumination Microscopy (SSIM)

SSIM, in principle, provides fluorescence imaging with unlimited resolution. In theory, by using a saturated excitation illumination, the excited-state of fluorophore is depleted to the ground state to generate a sinusoidal emission pattern and recorded on an area-array CCD detector. Due to the nonlinear effect (Gustafsson, 2005), extremely high-resolution can be achieved which is then reconstructed to generate super-resolution image (Huang et al., 2009). Because of this nonlinear fluorescence effect, there is severe quenching, bleaching, and saturation phenomena which ultimately affects the signal-to-noise ratio, and resolution limit. A recent study used the photoswitchable fluorescent protein Dronpa (Ando et al., 2004), to generate a nonlinear response from the sample using a wide-field total internal reflection fluorescence (TIRF) structured-illumination setup and resolved information in two dimensions at four times the conventional diffraction limit and validated (Rego et al., 2012).

4.2.4. Stimulated-Emission-Depletion Fluorescence Microscopy

Another more dynamic approach to resolve diffraction resolution limitations, is based on the saturated stimulated emission using two synchronized ultrafast laser sources (Hell and Wichmann, 1994; Klar et al., 2000; Westphal and Hell, 2005). STED is a scanning fluorescence microscopy which can resolve up to 35 nm by employing stimulated emission to inhibit the fluorescence process in the outer regions of the excitation point-spread function. Additional advantage of this system is that it can produce three-dimensional images of fluorescent objects owing to its far field scanning nature, as opposed to the near filed scanner principle of CLSM or TPM. Live cell STED microscopy has been successfully applied to the imaging of gap junctions and vimentin (Hein et al., 2010), actin (Fitzpatrick et al., 2009), and synaptic vesicles (Westphal et al., 2008). Using modified artificial CSF to potentiate the synapses i.e., chemical LTP and STED microscopy to image the resulting induced morphological plasticity of dendritic spines, Nägerl and his group reported structural changes in the 40–60% of the spines following the stimulation, while fewer than 10% of the spines appeared plastic under unstimulated conditions. Observed changes were also validated with local tetanic stimulation of afferent fibers (Nägerl et al., 2008). Because of the STED's potential to resolve fine spine structures, authors were able to report plastic changes in size and shape of the spines. For example, round or bulbous spine heads transformed to cup-like structure and even split into two heads. STED microscopy has revealed that synaptotagmin remains clustered after synaptic vesicle exocytosis (Willig et al., 2006).

CONCLUSION

This chapter highlighted the principles and applications of techniques-electrophysiology, optogenetics, optical markers and imaging tools, which have significantly contributed to our understanding of synaptic structure and plasticity mechanisms. In combination with the recent advances in the biochemical and genetic approaches to probe the roles of specific genes and proteins involved in synaptic plasticity, these tools have provided investigators an opportunity to study neuronal morphology and physiology at single circuit or single synapse level in a spatio-temporally regulated manner. Progressive and revolutionary advances in such techniques will push further our ability to study activity-dependent changes in neuronal signaling with high spatiotemporal resolution, and offer new insights into the basics of synaptic plasticity. This will eventually improve the understanding of the intricate electrical and chemical machineries of the brain and alterations in the synaptic plasticity associated with disorders and will help to find better prognosis as well as therapeutic targets.

REFERENCES

Abrahamsson T, Lalanne, Watt AJ, Sjöström PJ (2016) *In Vitro* Investigation of Synaptic Plasticity. *Cold Spring Harbor protoc* 2016: 087262.

Adamantidis AR, Tsai HC, Boutrel B, Zhang F, Stuber GD, Budygin EA, Touriño C, Bonci, A, Deisseroth K, de Lecea L (2011) Optogenetic interrogation of dopaminergic modulation of the multiple phases of reward-seeking behavior. *The Journal of Neuroscience: J Neurosci* 31: 10829-10835.

Ando R, Mizuno H, Miyawaki A (2004) Regulated fast nucleocytoplasmic shuttling observed by reversible protein highlighting. *Science* 306: 1370-1373.

Arendt T (2001) Disturbance of neuronal plasticity is a critical pathogenetic event in Alzheimer's disease. *Int J Dev Neurosci* 19: 231-245.

Arendt T, Brückner MK, Bigl V (1991) Maintenance of neuronal plasticity in the reticular core and changes in trophic activity in Alzheimer's disease. *Ann N Y Acad Sci.* 640: 210-214.

Bayazitov IT, Richardson RJ, Fricke RG, Zakharenko SS (2007) Slow presynaptic and fast postsynaptic components of compound long-term potentiation. *J Neurosci* 27: 11510-11521.

Ben-Simon Y, Rodenas-Ruano A, Alviña K, Lam AD, Stuenkel EL, Castillo PE, Ashery U (2015) A Combined Optogenetic-Knockdown Strategy Reveals a Major Role of Tomosyn in Mossy Fiber Synaptic Plasticity. *Cell Rep* 12: 396-404.

Berberich S, Pohle J, Pollard M, Barroso-Flores J, Köhr G (2017) Interplay between global and pathway-specific synaptic plasticity in CA1 pyramidal cells. *Sci Rep* 7: 17040.

Betz WJ, Bewick GS (1992) Optical analysis of synaptic vesicle recycling at the frog neuromuscular junction. *Science* 255: 200-203.

Betzig E, Patterson GH, Sougrat R, Lindwasser OW, Olenych S, Bonifacino JS, Davidson MW, Lippincott-Schwartz J, Hess HF (2006) Imaging intracellular fluorescent proteins at nanometer resolution. *Science* 313: 1642-1645.

Bliss TV, Collingridge GL (1993) A synaptic model of memory: long-term potentiation in the hippocampus. *Nature* 361: 31-39.

Bliss TV, Lomo T (1973) Long-lasting potentiation of synaptic transmission in the dentate area of the anaesthetized rabbit following stimulation of the perforant path. *J Physiol* 232: 331-356.

Blundon JA, Zakharenko SS (2008) Dissecting the components of long-term potentiation. *Neuroscientist* 14: 598-608.

Boyden, ES, Zhang F, Bamberg E, Nagel G, Deisseroth K (2005) Millisecond-timescale, genetically targeted optical control of neural activity. *Nat Neurosci* 8: 1263-1268.

Chater TE, Henley JM, Brown JT, Randall AD (2010) Voltage- and temperature-dependent gating of heterologously expressed channelrhodopsin-2. *J Neurosci Methods* 193: 7-13.

Chaudhury S, Jain S, Wadhwa S (2010) Expression of synaptic proteins in the hippocampus and spatial learning in chicks following prenatal auditory stimulation. *Dev neurosci.* 32(2): 114-124.

Choquet D (2010) Fast AMPAR trafficking for a high-frequency synaptic transmission. *Eur J Neurosci* 32: 250-260.

Chun S, Bayazitov IT, Blundon JA, Zakharenko SS (2013) Thalamocortical LTP becomes gated after the early critical period in the auditory cortex. *J Neurosci* 33: 7345-7357.

Cirrito JR, Yamada KA, Finn MB, Sloviter RS, Bales KR, May PC, Schoepp DD, Paul SM, Mennerick S, Holtzman DM (2005) Synaptic activity regulates interstitial fluid amyloid-beta levels *in vivo*. *Neuron* 48: 913-922.

Cole KS (1949) Dynamic electrical characteristics of the squid axon membrane. *Arch Sci Physiol* 3: 253–258.

Collins CA, DiAntonio A (2007) Synaptic development: insights from Drosophila. *Curr Opin Neurobiol* 17: 35-42.

Conforti L (2012) Chapter 20 - Patch-Clamp Techniques. In *Cell Physiology Source Book* (Fourth Edition), N. Sperelakis, ed. (San Diego: Academic Press): 369-381.

Cottrell JR, Dubé GR, Egles C, Liu G (2000) Distribution, density, and clustering of functional glutamate receptors before and after synaptogenesis in hippocampal neurons. *J Neurophysiol* 84: 1573-1587.

Cui, Z., Lindl, K. A., Mei, B., Zhang, S. and Tsien, J. Z. (2005), Requirement of NMDA receptor reactivation for consolidation and storage of nondeclarative taste memory revealed by inducible NR1 knockout. *Eur J Neurosci*, 22: 755–763.

Dahan M, Lévi S, Luccardini C, Rostaing P, Riveau B, Triller A (2003) Diffusion Dynamics of Glycine Receptors Revealed by Single-Quantum Dot Tracking. *Science* 302: 442-445.

Dani A, Huang B, Bergan J, Dulac C, Zhuang X (2010) Super-resolution Imaging of Chemical Synapses in the Brain. *Neuron* 68: 843-856.

Ding JB, Takasaki KT, Sabatini BL (2009) Supraresolution imaging in brain slices using stimulated-emission depletion two-photon laser scanning microscopy. *Neuron* 63: 429-437.

Egger MD, Petrăn M (1967) New reflected-light microscope for viewing unstained brain and ganglion cells. *Science* 157: 305-307.

Emptage N, Bliss TV, Fine A (1999) Single synaptic events evoke NMDA receptor-mediated release of calcium from internal stores in hippocampal dendritic spines. *Neuron* 22: 115-124.

Emptage NJ, Reid CA, Fine A, Bliss TVP (2003) Optical quantal analysis reveals a presynaptic component of LTP at hippocampal Schaffer-associational synapses. *Neuron* 38: 797-804.

Enoki R, Hu YL, Hamilton D, Fine A (2009) Expression of long-term plasticity at individual synapses in hippocampus is graded, bidirectional, and mainly presynaptic: optical quantal analysis. *Neuron* 62: 242-253.

Fitzpatrick JAJ, Yan Q, Sieber JJ, Dyba M, Schwarz U, Szent-Gyorgyi C, Woolford CA, Berget PB, Waggoner AS, Bruchez MP (2009) STED nanoscopy in living cells using Fluorogen Activating Proteins. *Bioconjug Chem* 20: 1843-1847.

Goshen I, Brodsky M, Prakash R, Wallace J, Gradinaru V, Ramakrishnan C, Deisseroth K (2011) Dynamics of retrieval strategies for remote memories. *Cell* 147: 678-689.

Gradinaru V, Zhang F, Ramakrishnan C, Mattis J, Prakash R, Diester I, Goshen I, Thompson KR, Deisseroth K (2010) Molecular and cellular approaches for diversifying and extending optogenetics. *Cell* 141: 154-165.

Gu Z, Yakel JL (2011) Timing-dependent septal cholinergic induction of dynamic hippocampal synaptic plasticity. *Neuron* 71: 155-165.

Grutzendler J, Kasthuri N, Gan WB (2002) Long-term dendritic spine stability in the adult cortex. *Nature* 420:812–816.

Gustafsson MGL (2005) Nonlinear structured-illumination microscopy: Wide-field fluorescence imaging with theoretically unlimited resolution. *Proc Nat Acad Sci USA* 102: 13081-13086.

Hamill OP, Marty A, Neher E, Sakmann B, Sigworth FJ (1981) Improved patch-clamp techniques for high-resolution current recording from cells and cell-free membrane patches. *Pflugers Archiv* 391: 85-100.

Harvey CD, Ehrhardt AG, Cellurale C, Zhong H, Yasuda R, Davis RJ, Svoboda K (2008) A genetically encoded fluorescent sensor of ERK activity. *Proc Nat Acad Sci USA* 105, 19264-19269.

Hein B, Willig KI, Wurm CA, Westphal V, Jakobs S, Hell SW (2010) Stimulated emission depletion nanoscopy of living cells using SNAP-tag fusion proteins. *Biophys J* 98: 158-163.

Hell SW, Wichmann J (1994) Breaking the diffraction resolution limit by stimulated emission: stimulated-emission-depletion fluorescence microscopy. *Opt Lett* 19: 780-782.

Helmchen F, DenkW (2005) Deep tissue two-photon microscopy. *Nat Methods* 2: 932-940.

Hosokawa T, Bliss TV, Fine A (1994) Quantitative three-dimensional confocal microscopy of synaptic structures in living brain tissue. *Microsc Res Tech.* 29: 290-296.

Huang B, Bates M, Zhuang X (2009) Super resolution fluorescence microscopy. *Ann Rev Biochem* 78: 993-1016.

Huff ML, Miller RL, Deisseroth K, Moorman DE, LaLumiere RT (2013) Posttraining optogenetic manipulations of basolateral amygdala activity modulate consolidation of inhibitory avoidance memory in rats. *Proc Nat Acad Sci USA* 110: 3597-3602.

Johansen JP, Hamanaka H, Monfils MH, Behnia R, Deisseroth K, Blair HT, LeDoux JE (2010) Optical activation of lateral amygdala pyramidal cells instructs associative fear learning. *Proc Nat Acad Sci USA* 107: 12692-12697.

Johnston D, Hopkins WF, Gray R (1988) Norepinephrine Enhances Long-Term Potentiation at Hippocampal Mossy Fiber Synapses. In *Synaptic Plasticity in the Hippocampus* (Springer, Berlin, Heidelberg): 57-60.

Kerchner GA, Nicoll RA (2008) Silent synapses and the emergence of a postsynaptic mechanism for LTP. *Nat Rev Neurosci* 9: 813-825.

Kirkwood A, Lee HK, Bear MF (1995) Co-regulation of long-term potentiation and experience-dependent synaptic plasticity in visual cortex by age and experience. *Nature* 375: 328-331.

Klar TA, Jakobs S, Dyba M, Egner A, Hell SW (2000) Fluorescence microscopy with diffraction resolution barrier broken by stimulated emission. *Proc Nat Acad Sci USA* 97: 8206-8210.

Kornreich BG (2007) The patch clamp technique: Principles and technical considerations. *J Vet Cardiol* 9: 25-37.

Kumar V, nag TC, Sharma U, Jagannathan NR, Wadhwa S (2014) Differential effects of prenatal chronic high-decibel noise and music exposure on the excitatory and inhibitory synaptic components of the auditory cortex analog in developing chicks (Gallus gallus domesticus). *Neurosci* 269: 302-317.

Lakowicz JR (2006) *Principles of Fluorescence Spectroscopy* (Springer).

Lee D, Lee AK (2017) *In Vivo* Patch-Clamp Recording in Awake Head-Fixed Rodents. *Cold Spring Harb Protoc* 2017: pdb.prot095802.

Lee SJR., Escobedo-Lozoya Y, Szatmari EM, Yasuda R (2009) Activation of CaMKII in single dendritic spines during long-term potentiation. *Nature* 458: 299-304.

Lynch G, Gall CM (2013) Mechanism Based Approaches for Rescuing and Enhancing Cognition. *Front Neurosci* 7: 143.

Lynch G, Larson J, Kelso S, Barrionuevo G, Schottler F (1983) Intracellular injections of EGTA block induction of hippocampal long-term potentiation. *Nature* 305: 719-721

Lynch G, Palmer LC, Gall CM (2011) The likelihood of cognitive enhancement. *Pharmacol Biochem Behav*. 99: 116-129.

Makino H, Malinow R (2009) AMPA receptor incorporation into synapses during LTP: the role of lateral movement and exocytosis. *Neuron* 64: 381-390.

Malenka RC, Kauer JA, Zucker RS, Nicoll RA (1988) Postsynaptic calcium is sufficient for potentiation of hippocampal synaptic transmission. *Science* 242: 81-84.

Man HY, Ju W, Ahmadian G, Wang YT (2000) Intracellular trafficking of AMPA receptors in synaptic plasticity. *Cell Mol Life Sci* 57: 1526-1534.

Manabe T (2012) Analysis of Synaptic Plasticity with the Slice Patch-Clamp Recording Technique. In *Patch Clamp Techniques* (Springer, Tokyo): 147-157.

Manabe T, Aiba A, Yamada A, Ichise T, Sakagami H, Kondo H, Katsuki M (2000) Regulation of Long-Term Potentiation by H-Ras through NMDA Receptor Phosphorylation. *J Neurosci* 20: 2504-2511.

Miyawaki A (2003) Visualization of the Spatial and Temporal Dynamics of Intracellular Signaling. *Dev Cell* 4: 295-305.

Miyawaki A., Llopis, J., Heim, R., McCaffery, J.M., Adams, J.A., Ikura, M., and Tsien, R.Y. (1997). Fluorescent indicators for Ca2+ based on green fluorescent proteins and calmodulin. *Nature* 388: 882-887.

Murakoshi H, Lee SJ, Yasuda R (2008) Highly sensitive and quantitative FRET-FLIM imaging in single dendritic spines using improved non-radiative YFP. *Brain Cell Biol* 36: 31-42.

Murakoshi H, Shin ME, Parra-Bueno P, Szatmari EM, Shibata ACE, Yasuda R (2017) Kinetics of Endogenous CaMKII Required for Synaptic Plasticity Revealed by Optogenetic Kinase Inhibitor. *Neuron* 94: 690.

Murakoshi H, Wang H, Yasuda R (2011) Local, persistent activation of Rho GTPases during plasticity of single dendritic spines. *Nature* 472: 100-104.

Murthy VN, Sejnowski TJ, Stevens CF (1997) Heterogeneous release properties of visualized individual hippocampal synapses. *Neuron* 18: 599-612.

Nägerl UV, Eberhorn N, Cambridge SB, Bonhoeffer T (2004) Bidirectional Activity-Dependent Morphological Plasticity in Hippocampal Neurons. *Neuron* 44: 759-767.

Nägerl UV, Willig KI, Hein B, Hell SW, Bonhoeffer T (2008) Live-cell imaging of dendritic spines by STED microscopy. *Proc Nat Acad Sci USA* 105: 18982-18987.

Nisticò R, Pignatelli M, Piccinin S, Mercuri NB, Collingridge G (2012) Targeting Synaptic Dysfunction in Alzheimer's Disease Therapy. *Mol Neurobiol* 46: 572-587.

Okamoto KI, Nagai T, Miyawaki A, Hayashi Y (2004) Rapid and persistent modulation of actin dynamics regulates postsynaptic reorganization underlying bidirectional plasticity. *Nat Neurosci* 7: 1104-1112.

Olsen DP, Keshishian H (2012) Experimental Methods for Examining Synaptic Plasticity in Drosophila. *Cold Spring Harbor Protoc* 2012: 226-230.

Pascoli V, Turiault M, Lüscher C (2011) Reversal of cocaine-evoked synaptic potentiation resets drug-induced adaptive behaviour. *Nature* 481: 71-75.

Piccolino M (1998) Animal electricity and the birth of electrophysiology: the legacy of Luigi Galvani. *Brain Res Bull* 46: 381-407.

Rahamimoff R, Melamed N (1993) Visualization of synaptic structure and function with confocal microscopy: calcium fluctuations and oscillations. *Neurosci Res* 16: 173-180.

Ramirez S, Liu X, Lin PA, Suh J, Pignatelli M, Redond, RL, Ryan TJ Tonegawa S (2013) Creating a false memory in the hippocampus. *Science* 341: 387-391.

Rego EH, Shao L, Macklin JJ, Winoto L, Johansson GA, Kamps-Hughes N, Davidson MW, Gustafsson MGL (2012) Nonlinear structured-illumination microscopy with a photoswitchable protein reveals cellular structures at 50-nm resolution. *Proc Nat Acad Sci USA* 109: E135-E143.

Reid CA, Dixon DB, Takahashi M, Bliss TVP, Fine A (2004) Optical quantal analysis indicates that long-term potentiation at single hippocampal mossy fiber synapses is expressed through increased release probability, recruitment of new release sites, and activation of silent synapses. *The Journal of Neuroscience: J Neurosci* 24: 3618-3626.

Reid CA, Fabian-Fine R, Fine A (2001) Postsynaptic calcium transients evoked by activation of individual hippocampal mossy fiber synapses. *The Journal of Neuroscience: J Neurosci* 21: 2206-2214.

Robitaille R, Adler EM, Charlton MP (1990) Strategic location of calcium channels at transmitter release sites of frog neuromuscular synapses. *Neuron* 5: 773-779.

Rolls A, Colas D, Adamantidis A, Carter M, Lanre-Amos T, Heller HC, de Lecea, L (2011) Optogenetic disruption of sleep continuity impairs memory consolidation. *Proc Nat Acad Sci USA* 108: 13305-13310.

Rust MJ, Bates M, Zhuang X (2006) Sub-diffraction-limit imaging by stochastic optical reconstruction microscopy (STORM*). Nat Methods* 3: 793-795.

Ryan TA, Reuter H, Wendland B, Schweizer FE, Tsien RW, Smith SJ (1993). The kinetics of synaptic vesicle recycling measured at single presynaptic boutons. *Neuron* 11: 713-724.

Sanyal T, Kumar V, Nag TC, Jain S, Sreenivas V, Wadhwa S (2013) Prenatal loud music and noise: differential impact on physiological arousal, hippocampal synaptogenesis and spatial behavior in one day-old chicks. *PLoS One* 8(7): e67347.

Sasaki T, Beppu K, Tanaka KF, Fukazawa Y, Shigemoto R, Matsui K (2012) Application of an optogenetic byway for perturbing neuronal activity via glial photostimulation. *Proc Nat Acad Sci USA* 109: 20720-20725.

Schultheis C, Liewald JF, Bamberg E, Nagel G, Gottschalk A (2011) Optogenetic long-term manipulation of behavior and animal development. *PloS One* 6: e18766.

Shinohara Y, Hirase H, Watanabe M, Itakura M, Takahashi M, Shigemoto R (2008) Left-right asymmetry of the hippocampal synapses with differential subunit allocation of glutamate receptors. *Proc Nat Acad Sci USA* 105: 19498-19503.

Shroff H, Galbraith CG, Galbraith JA, Betzig E (2008) Live-cell photoactivated localization microscopy of nanoscale adhesion dynamics. *Nat Methods* 5, 417-423.

Takao K, Okamoto KI, Nakagawa T, Neve RL, Nagai T, Miyawaki A, Hashikawa T, Kobayashi S, Hayashi Y (2005) Visualization of synaptic Ca2+ /calmodulin-dependent protein kinase II activity in living neurons. *J Neurosci* 25: 3107-3112.

Tardin C, Cognet L, Bats C, Lounis B, Choquet D (2003) Direct imaging of lateral movements of AMPA receptors inside synapses. *The EMBO J* 22: 4656-4665.

Tian L, Hires SA, Mao T, Huber D, Chiappe ME, Chalasani SH, Petreanu L, Akerboom J, McKinney SA, Schreiter ER, Bargmann CI, Jayaraman V, Svoboda K, Looger LL (2009) Imaging neural activity in worms, flies and mice with improved GCaMP calcium indicators. *Nat Methods* 6: 875-881.

Tong G, Malenka RC, Nicoll RA (1996) Long-Term Potentiation in Cultures of Single Hippocampal Granule Cells: A Presynaptic Form of Plasticity. *Neuron* 16: 1147-1157.

Triller A, Choquet D (2005) Surface trafficking of receptors between synaptic and extrasynaptic membranes: and yet they do move! *Trends Neurosci* 28: 133-139.

Tsai J, Grutzendler J, Duff K, Gan WB (2004) Fibrillar amyloid deposition leads to local synaptic abnormalities and breakage of neuronal branches. *Nat Neurosci* 7:1181–1183.

Urban BE, Xiao L, Dong B, Chen S, Kozorovitskiy Y, Zhang HF (2017) Imaging neuronal structure dynamics using 2-photon super-resolution patterned excitation reconstruction microscopy. *Journal Biophotonics* (in press).

Vanderklish PW, Krushel LA, Holst BH, Gally JA, Crossin KL, Edelman GM (2000) Marking synaptic activity in dendritic spines with a calpain substrate exhibiting fluorescence resonance energy transfer. *Proc Nat Acad Sci USA* 97: 2253-2258.

Wang YL, Taylor DL (1981) Probing the dynamic equilibrium of actin polymerization by fluorescence energy transfer. *Cell* 27: 429-436.

Ward B, McGuinness L, Akerman CJ, Fine A, Bliss TVP, Emptage NJ (2006) State-dependent mechanisms of LTP expression revealed by optical quantal analysis. *Neuron* 52: 649-661.

Weiss S (1999). Fluorescence Spectroscopy of Single Biomolecules. *Science* 283: 1676-1683.

Westphal V, Hell SW (2005) Nanoscale resolution in the focal plane of an optical microscope. *Phys Rev Lett* 94: 143903.

Westphal V, Rizzoli SO, Lauterbach MA, Kamin D, Jahn R, Hell SW (2008) Video-Rate Far-Field Optical Nanoscopy Dissects Synaptic Vesicle Movement. *Science* 320: 246-249.

Willig KI, Rizzoli SO, Westphal V, Jahn R, Hell SW (2006) STED microscopy reveals that synaptotagmin remains clustered after synaptic vesicle exocytosis. *Nature* 440: 935-939.

Xiong W, Jin X (2012). Optogenetic field potential recording in cortical slices. *J Neurosci Methods* 210: 119-124.

Yasuda R, Harvey CD, Zhong H, Sobczyk A, van Aelst L, Svoboda K (2006) Supersensitive Ras activation in dendrites and spines revealed by two-photon fluorescence lifetime imaging. *Nat Neurosci* 9: 283-291.

Zakharenko SS, Zablow L, Siegelbaum SA (2001) Visualization of changes in presynaptic function during long-term synaptic plasticity. *Nat Neurosci* 4: 711-717.

Zhang F, Gradinaru V, Adamantidis AR, Durand R, Airan RD, de Lecea L, and Deisseroth K (2010) Optogenetic interrogation of neural circuits: technology for probing mammalian brain structures. *Nature Protoc* 5: 439-456.

Zhang F, Wang LP, Boyden ES, Deisseroth K (2006) Channelrhodopsin-2 and optical control of excitable cells. *Nat Method*s 3: 785-792.

Zhang YP, Oertner, TG (2007) Optical induction of synaptic plasticity using a light-sensitive channel. *Nat Methods* 4: 139-141.

Zuo Y, Yang G, Kwon E, Gan WB (2005) Long-term sensory deprivation prevents dendritic spine loss in primary somatosensory cortex. *Nature* 436:261–265.

Zhao M, Choi YS, Obrietan K, Dudek SM (2007) Synaptic Plasticity (and the Lack Thereof) in Hippocampal CA2 Neurons. *J Neurosci* 27: 12025-12032.

Zhao XY, Liu M.G, Yuan DL, Wang Y, He Y, Wang DD, Chen XF. Zhang FK, Li H, He XS, Chen J (2009) Nociception-induced spatial and temporal plasticity of synaptic connection and function in the hippocampal formation of rats: a multi-electrode array recording. *Mol Pain* 5: 55.

About the Editor

Sraboni Chaudhury is currently a Research Scientist at Molecular Behavior Neuroscience Institute, University of Michigan, USA. She has finished her Master in Neuroscience in 2003 from Jiwaji University Gwalior, India and PhD in Neurobiology in 2009 from Department of Anatomy, All India Institute of Medical Sciences, New Delhi, India. She did her postdoctoral research at University of Texas Health Science Center at San Antonio, Texas, USA and University of Michigan, USA. Her PhD work deals with role of prenatal environmental stimuli in the development of Hippocampus which is the genesis of fundamental of this book. Her work during PhD and postdoctoral research revolves in understanding the basis of plasticity of brain during development, adulthood and in diseased condition. She has published several article in national and international journal including Proceedings of the National Academy of Sciences of the United States of America and Journal of Neuroscience (Official Journal of Society of Neuroscience). She is serving as reviewer as well as editorial board member to several journals. She is also an Associate Editor of Degenerative Intellectual & Developmental Disabilities Journal.

INDEX

A

B

D

E

F

G

H

I

K

L

M

N

O

P

R

S

X

A